Dynamic Fracture

Elsevier Internet Homepage-http://www.elsevier.com
Consult the Elsevier homepage for full catalogue information on all books, journals and electronic products and services.

Elsevier Titles of Related Interest

ALLIX and HILD
Continuum Damage Mechanics of Materials and Structures.
ISBN: 008-043918-7

BLACKMAN *ET AL.*
Fracture of Polymers, Composites and Adhesives II
ISBN: 008-044195-9

CARPINTERI *ET AL.*
Biaxial/Multiaxial Fatigue and Fracture.
ISBN: 008-044129-7

ELICES and LLORCA
Fiber Fracture.
ISBN: 008-044104-1

FRANÇOIS and PINEAU
From Charpy to Present Impact Testing.
ISBN: 008-043970-5

FUENTES *ET AL.*
Fracture Mechanics: Applications and Challenges.
ISBN: 008-043699-4

JONES
Failure Analysis Case Studies II.
ISBN: 008-043959-4

JONES
Failure Analysis Case Studies III.
ISBN: 008-044447-4

MILNE *ET AL.*
Comprehensive Structural Integrity.
ISBN: 008-043749-4

MOORE
The Application of Fracture Mechanics to Polymers,
Adhesives and Composites
ISBN: 008-044205-6

MOORE *ET AL.*
Fracture Mechanics Testing Methods for Polymers,
Adhesives and Composites.
ISBN: 008-043689-7

MURAKAMI
Metal Fatigue Effects of Small Defects and Nonmetallic
Inclusions.
ISBN: 008-044064-9

PAULINO
Fracture of Functionally Graded Materials.
ISBN: 008-044160-2

K.S. RAVICHANDRAN *ET AL.*
Small Fatigue Cracks: Mechanics, Mechanisms &
Applications.
ISBN: 008-043011-2

RÉMY and PETIT
Temperature-Fatigue Interaction.
ISBN: 008-043982-9

TANAKA & DULIKRAVICH
Inverse Problems in Engineering Mechanics IV.
ISBN: 008-044268-4

WILLIAMS & PAVAN
Fracture of Polymers, Composites and Adhesives.
ISBN: 008-043710-9

Related Journals
Free specimen copy gladly sent on request. Elsevier Ltd, The Boulevard, Langford Lane, Kidlington, Oxford, OX5 1GB, UK

Acta Metallurgica et Materialia
Cement and Concrete Research
Composite Structures
Computers and Structures
Corrosion Science
Engineering Failure Analysis
Engineering Fracture Mechanics
European Journal of Mechanics A & B
International Journal of Fatigue
International Journal of Impact Engineering
International Journal of Mechanical Sciences
International Journal of Non-Linear Mechanics
International Journal of Plasticity

International Journal of Pressure Vessels & Piping
International Journal of Solids and Structures
Journal of Applied Mathematics and Mechanics
Journal of Construction Steel Research
Journal of the Mechanics and Physics of Solids
Materials Research Bulletin
Mechanics of Materials
Mechanics Research Communications
NDT&E International
Scripta Metallurgica et Materialia
Theoretical and Applied Fracture Mechanics
Tribology International
Wear

To Contact the Publisher
Elsevier Science welcomes enquiries concerning publishing proposals: books, journal special issues, conference proceedings, etc.
All formats and media can be considered. Should you have a publishing proposal you wish to discuss, please contact, without
obligation, the publisher responsible for Elsevier's mechanics and structural integrity publishing programme:

Dean Eastbury
Senior Publishing Editor, Materials Science & Engineering
Elsevier Ltd
The Boulevard, Langford Lane   Phone: +44 1865 843580
Kidlington, Oxford   Fax: +44 1865 843920
OX5 1GB, UK   E.mail: d.eastbury@elsevier.com

General enquiries, including placing orders, should be directed to Elsevier's Regional Sales Offices - please access the Elsevier
homepage for full contact details (homepage details at the top of this page).

Dynamic Fracture

K. Ravi-Chandar

Department of Aerospace Engineering and Engineering Mechanics
The University of Texas, Austin, USA

2004

ELSEVIER

Amsterdam – Boston – Heidelberg – London – New York – Oxford – Paris
San Diego – San Francisco – Singapore – Sydney – Tokyo

ELSEVIER B.V.
Sara Burgerhartstraat 25
P.O. Box 211, 1000 AE Amsterdam
The Netherlands

ELSEVIER Inc.
525 B Street, Suite 1900
San Diego, CA 92101-4495
USA

ELSEVIER Ltd
The Boulevard, Langford Lane
Kidlington, Oxford OX5 1GB
UK

ELSEVIER Ltd
84 Theobalds Road
London WC1X 8RR
UK

First edition 2004

Library of Congress Cataloging in Publication Data
A catalog record is available from the Library of Congress.

British Library Cataloguing in Publication Data
A catalogue record is available from the British Library.

ISBN: 0-08-044352-4

∞ The paper used in this publication meets the requirements of ANSI/NISO Z39.48-1992 (Permanence of Paper).
Printed in The Netherlands.

Preface

The aim of this book is to provide an overview of dynamic fracture in nominally brittle materials. Brittle fracture in solids has attracted much attention over the second half of the 20th century from engineers as well as physicists due both to its technological interest and inherent scientific curiosity. Early investigations into brittle fracture were performed under dynamic loads without the presence of a dominant crack in the body; this type of loading causes spalling of the material through the nucleation, growth and coalescence of multiple cracks in the spall plane. The advent of a fracture mechanical description of material failure that began with the pioneering effort of Griffith has steered recent dynamic brittle fracture investigations away from the spall problem and towards a crack-dominated approach; this book focuses on the crack problem.

The literature on the subject is vast; a quick search on databases with keywords "dynamic and fracture" yields results numbering in many thousands; and research in the field continues at a significant pace, especially in the area of numerical simulations. Coverage of the topic must therefore be selective. I have restricted attention for the most part to opening mode cracks in homogeneous nominally brittle materials; extensions to shearing mode cracks can be made easily by applying the criterion of local symmetry insisting that the crack choose a path so as to ensure a locally opening mode crack. Such a claim is not valid when considering interfaces or graded materials; there are many investigations of this problem, but these are not considered in this book. I have also chosen to describe the classical elastodynamic interpretation of the problem in detail and provide only glimpses of new approaches such as the discrete models and cohesive zone models; while such models are potentially powerful in providing detailed numerical simulations of dynamic fracture problems, fundamental considerations regarding the determination of appropriate material properties for inclusion in these models and the quantitative comparison of the results of the simulations to experimental observations remain open issues. Dynamic failure criteria and the limitations in their applicability in assessing structural integrity are the primary focus of this book. Analytical characterization of the crack tip state, methods of implementation of dynamic fracture experiments, diagnostic techniques and their limitations, and interpretation of dynamic failure criteria are discussed in great detail.

The introductory chapter provides examples of problems where structural integrity assessment through the elastodynamic fracture theory is important; an outline of the topics covered is also presented there. This book should be accessible to graduate students with a background in solid mechanics. This book could be used as a resource in courses devoted to fracture mechanics or experimental mechanics. This book should also be useful to workers in the field of structural integrity assessment. For example, implementation of test methodologies for determination of rate-dependent dynamic crack initiation toughness or crack growth criterion discussed in this book should be a simple task given modern test and measurement equipment and data acquisition systems.

I am deeply indebted to the many contributors to this field for numerous discussions that helped me in understanding the mechanics and mechanisms of dynamic fracture. It is indeed my pleasure to specifically acknowledge Wolfgang Knauss (California Institute of Technology) for support and guidance during my graduate studies in this field and for his continued encouragement and friendship through all these years. I am also indebted to Wolfgang Knauss, Michael Marder (University of Texas at Austin), and Sridhar Krishnaswamy (Northwestern University) for reading drafts of all or portions of the manuscript.

I thank Dean Eastbury, Senior Publisher, for encouraging me to pursue this book. I also thank Sharon Brown and Carol Cooper who coordinated the production of the book, for putting up with the many delays on my part.

I thank my wife Hema and son Prakash, for their love, support and the many sacrifices they endured as I worked on this book. I dedicate this book to my parents, Padma and Sunder Krishnaswamy.

K. Ravi-Chandar
Austin, Texas
2004

Contents

Chapter 1

Introduction

Rapidly applied loads are encountered in a number of applications. In some cases such loads might be applied deliberately, as for example in problems of blasting, mining, and comminution or fragmentation; in other cases, such dynamic loads might arise from accidental conditions. Regardless of the origin of the rapid loading, it is necessary to understand the mechanisms and mechanics of fracture under dynamic loading conditions in order to design suitable procedures for assessing the susceptibility to fracture. Quite apart from its repercussions in the area of structural integrity, fundamental scientific curiosity has continued to play a large role in engendering interest in dynamic fracture problems.

At the outset, it is essential to identify the range of loading rates in which the dynamic analysis based on wave propagation is important. We illustrate this with a simple example: consider a single-edge notched specimen illustrated in Fig. 1.1. Let the load be applied by a tup, falling at a speed v and impacting on the specimen. The impact generates a stress wave that travels into the specimen, interacts with the crack and reflects from the far end. The time scale of the first interaction is w/C_{d}, where w is the depth of the specimen and C_{d} is the longitudinal wave speed of the material. If the impact speed v is sufficiently high, it is possible to initiate the fracture event before the arrival of the stress waves at the supporting posts—i.e. at times $t < l/C_{\mathrm{d}}$ and before the wave reflection from the bottom travels back up to the top $t < 2w/C_{\mathrm{d}}$. This is a truly transient dynamic condition and requires a full elastodynamic analysis of the problem. If the specimen does not break within this time scale, then the stress waves subsequently reflect between the top and bottom surfaces of the specimen and eventually put the beam into a vibratory motion at a frequency that corresponds to the natural frequency of the beam on its supports; note that this may occur only after several wave passes. In many impact tests, the measurements clearly indicate the arrival of different reflections. In cases where a steady vibratory state has been reached, it might be sufficient to perform quasi-static analysis, where the time-dependent load at the tup is measured and used in the static calculation of the stress state at the crack tip. At long times, these oscillations decay by transmission of the waves into the supports at the bottom end of the specimen and the full weight of the tup is held by the equilibrium bending stresses in the beam. The last stage corresponds to the quasi-static condition and is the realm of elastostatics. Of course, in practice, whether these regimes are exhibited or not

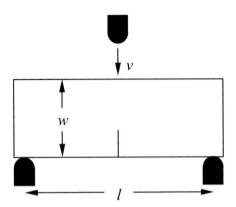

Figure 1.1 Geometry of an impact test configuration.

depends on the nature of the specimen and the material properties. There has been a lively debate in the literature, especially with respect to the pressurized thermal shock problem discussed in Section 1.1 about the suitability of a static analysis or the necessity for using a fully dynamic analysis. The interest in this book is on problems that fall primarily into the first group—fully elastodynamic problems. There are a number of examples of engineering structures that are subjected to dynamic loads and require a fully dynamic analysis. Here we describe a few examples to understand the topics considered in this book.

1.1 Pressurized Thermal Shock in Nuclear Containment Vessels

One of the major problems motivating investigations of dynamic fracture is in pressurized water reactors (PWR), where the containment vessels may be subjected to significant thermal shock loading during a loss of coolant accident (LOCA). These pressure vessels are assumed to contain crack-like flaws that result from various processing, handling and use conditions such as welding, load cycling, and stress-corrosion. The initial design of the pressure vessel according to the prevailing standards and codes usually accounts for the presence of such flaws. In case of a LOCA, a failure in the primary cooling system is compensated by emergency core coolant introduced into the inner wall of the vessel which drops the temperature of the hot vessel wall by more than 250°C; the vessel must be capable of sustaining this thermal shock (Cheverton et al., 1981). However, during operation of a PWR, additional concerns arise: first, the toughness of the material reduces with time as a result of radiation-induced damage, particularly along the inner wall. Second, in a LOCA, the steep thermal gradient generates large stresses, sufficient to cause growth of flaws present along the inner wall. Third, it is also possible that the thermal transient occurs while the pressure loading has not yet been released. Safe design of the vessel requires that this LOCA is a survivable incident, without propagation of a crack through the wall thickness. While initial designs were based on quasi-static linear elastic fracture mechanics analysis and material characterization, experiments have indicated that the event is truly dynamic and that multiple crack initiation, rapid crack propagation and crack arrest events can occur during the course of a single thermal shock event. Experiments have been conducted on thick-walled cylinders under thermal shock conditions (Cheverton et al., 1981) as well as in wide plates simulating the thermal shock in the pressure vessels (Pugh et al., 1988). Analyses of the wide-plate test results using quasi-static and dynamic fracture methodology were provided by Jung and Kanninen (1983), Bass et al. (1985), Brickstad and Nilsson (1986a,b), and Pugh et al. (1988). While there has been much debate over whether quasi-static analyses are conservative or not, consensus was reached on the need for the determination of the rate and temperature-dependent crack initiation and crack arrest properties.

1.2 Boiler and Pipeline Burst Problems

Pipelines transport oil and gas under high-pressure conditions over long distances. Millions of miles of such transmission and distribution pipelines exist around the world. Loss of revenue due to accidental rupture of these pipelines is in the order of tens of millions of dollars per year; in some instances, these accidents have led to loss of life as well. In the United States of America, according to the data collected by the Department of Transportation, in the last 25 years of the 20th century, such losses total about a billion dollars in revenue; in addition there have been too many fatalities. Pressurized boilers and pipelines are designed according to applicable codes of the American Society of Mechanical Engineers (ASME), American Petroleum Institute (API), American Gas Association (AGA), etc.; the failures observed in service are generally not due to poor or improper design, but mainly due to other causes: first, buried pipelines are ruptured due to incidental damage induced by digging or other operations in the vicinity of the pipeline; reports compiled by the US National Transportation Safety Board indicate that work crew installing utility lines for electric, cable and/or water supply near a gas line sometimes inadvertently damage gas distribution pipelines. Second, corrosion damage may have accumulated over time to the extent that the residual strength of the pipe falls below the operating stress levels; this is exacerbated by the fact that much of the infrastructure is aging and hence quite susceptible to failure. Repeated pressurization cycles experienced by the pipelines as they are pumped with different fluids for transmission along their length also induces fatigue cracks to grow to critical dimensions in some pipelines. Fractures or ruptures that initiate under such conditions may propagate at speeds that are a substantial fraction of the wave speed in the solid; cracks growing at speeds order of 500 m/s have been observed in steel pipes (Ives et al., 1974). Depressurization of the pipe as a result of escaping gases travels along the length of the pipe at the speed of sound in the medium, typically on the order of 300 m/s. Hence, the dynamically growing crack outruns the depressurization and the driving force for the crack is maintained over long distances. The loading conditions on these pipelines are quite complex; in addition to the internal pressure, the backfill provides an external loading. Furthermore, the leakage of the contents through the opening produced by the propagating fracture results in a decrease of loading that must be estimated through a coupled fluid–structure interaction problem.

A complete analysis of this problem requires consideration of the dynamics of all these interactions and has been attempted only under special conditions. The two main approaches to the problem have been empirical in nature; both rely upon burst tests and correlations between the burst energy and the Charpy impact energy.[1] The ASME Boiler and Pressure Vessel Code imposes a requirement of a minimum impact energy for the material and circumvents analysis of dynamic fracture. On the other hand, for gas pipeline applications, for example, the following empirical expression has been used to determine susceptibility to failure, once again correlated to experimental database (O'Donoghue et al., 1997):

$$\frac{2}{3}(C_v)_{\min} = 2.52 \times 10^{-4} R\sigma_h + 1.245 \times 10^{-5} \frac{Rh\sigma_h^2}{d}$$
$$- 0.627h - 6.8 \times 10^{-8} \frac{R^2 d}{h} \tag{1.1}$$

where $(C_v)_{\min}$ is the Charpy energy in joules, h the wall thickness, R the pipe radius, d the depth of the backfill, all in millimeters, and σ_h is the hoop stress in MPa. O'Donoghue et al. (1997) also show that expressions such as Eq. 1.1 are not useful outside the range of data used to obtain such empirical relationships and could be significantly nonconservative, pointing to the need for a fracture mechanics-based analysis even though this analysis is quite expensive. In spite of some lingering uncertainties in the dynamic fracture theory described in this book, it is capable of providing a good engineering estimate of rapid crack growth and arrest in such pressure vessel and pipeline applications.

1.3 Dynamic Fracture in Airplane Structures

Another class of problems that provide motivation for studies of dynamic loading arises in the evaluation of the integrity of airplane structures. Fuselages of airplanes are pressure vessels, designed to maintain a cabin pressure at a higher level than the ambient pressure, at the cruising altitude. Fatigue cracks emanate from regions of stress concentration and grow with repeated pressurization cycles associated with take-off and landing cycles. While the early examples of such fatigue cracks, such as the de Havilland Comet aircraft at the dawn of commercial jet aviation, resulted in catastrophic disintegration of the aircraft structure, structural design concepts have evolved to limit fatigue crack extension during the design lifetime of the structure. Recent accidents, such as the Aloha Airlines Boeing 737 aircraft incident in 1988, demonstrated that small cracks emanating from neighboring rivet holes can interact with each other and critical lengths sufficient to trigger dynamic crack growth can be reached. This phenomenon is referred to as multi-site damage. Growth of such cracks is similar to the problem of dynamic crack growth in the pipeline described above; at high crack speeds, the decompression may not appear rapidly enough to arrest the crack. Crack arrest or deflection methodologies based on the concept of tear straps are used here to limit the extent of crack growth; some of these concepts are

---

[1] The Charpy test and the impact energy are described further in Section 1.4.

discussed in Chapter 10. Once again, as in pipeline applications, such tear straps are currently designed based on empiricism relying on scale model and full-scale tests on actual structures rather than predictive models grounded in fracture mechanics. There has been some recent progress in evaluating the tear straps using principles of dynamic fracture mechanics (see Kosai and Kobayashi, 1991).

Additional motivation for analysis based on dynamic fracture mechanics is provided by examining the response of aircraft to blast. The terrorist attack on Pan Am Boeing 747 over Lockerbie, Scotland in 1988 resulted in significant loss of life; reconstruction of the damage to the aircraft indicated that the initial blast from the explosive charge resulted in small structural damage, but a large build-up of pressure in the fuselage, and further that the disintegration of the airplane resulted not from the initial blast, but due to dynamic growth of cracks triggered from the blast site by the high pressure still contained within in the fuselage (UK Air Accidents Investigation Branch, 1990). Fig. 1.2 shows the pattern of damage sustained by the aircraft; the initial blast resulted in the star-burst and petaling pattern marked as Region A. Major cracks, labeled Fractures 1–3 in the figure emanated from the star-burst and propagated along the length of the fuselage, not only from the blast over-pressure, but also likely due to cabin pressure. The turning of these cracks at the tear-straps is seen in Regions B, C, D and E. The investigators of the disintegration of PanAm 103 recommended that the survivability of aircraft to such small scale blasts may be increased by suitable design of the fuselage panels to inhibit rapid crack propagation over long lengths and by provision of panels specifically designed to break-off and vent the blast products and decrease the pressure quickly. Dynamic analysis of the fracture phenomena and appropriate characterization of material properties are essential for

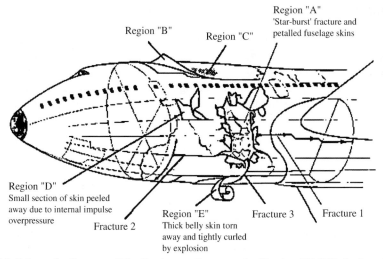

Figure 1.2 Schematic diagram of the fracture patterns on the PanAm 103 747 airplane that was destroyed over Lockerbie, Scotland by a terrorist attack. Region A identifies the initial blast damage; Fractures 1–3 are thought to have propagated as a result of the service pressure inside the cabin. Regions B–E illustrate the role of tear-straps in deflecting the cracks. (Reproduced from Aircraft Accident Report No 2/90 (EW/C 1094), UK Air Accidents Investigation Branch.)

successful incorporation of these ideas in design practice. Therefore, recent concerns of aircraft structural integrity motivated by the aging of the commercial and military aircraft fleet and by terrorist threats have focused some attention on the analysis of dynamic crack growth and arrest in airplanes.

1.4 Notched Bar Impact Testing of Metallic Materials

Impact resistance of materials has long been evaluated beginning with the Charpy test (Charpy, 1901). The data obtained from this and other similar tests are qualitative at best and not suitable for predictive analysis based on fracture mechanics. On the other hand, the test is quite easily performed and therefore quite useful for comparative ranking of different materials and in selection of materials for specific use. The ASTM E-23 standard presents a method for the determination of the energy absorbed during impact fracture in metallic materials. The basic idea behind the test is the following: a V-notched beam specimen is supported on an anvil and impacted by a mass moving with a sufficient energy; for a test to be valid under the standard, the impact speed must be in the range of 3–6 m/s. From a measure of the initial and final value of the potential energy of the mass, the energy absorbed in the fracture event, called the Charpy V-notched energy, C_v, is determined. Fig. 1.3 shows the temperature dependence of C_v typical of low-strength ferrous alloys. In these alloys C_v exhibits a remarkable transition from a brittle behavior at low temperatures (called the lower shelf) to a ductile behavior at high temperatures (called the upper shelf). The temperature at which the failure mechanism changes from brittle to ductile is called the nil-ductility temperature (NDT). Many variants of the Charpy test have been proposed; within the ASTM E-23 standard, there are other possible configurations for the impact test. The drop weight tear test (ASTM E-466), and the dynamic tear test (ASTM E-604, ASTM E-208) are other tests aimed at providing qualitative information regarding the nature of the fracture (brittle or ductile) and an estimate of the transition temperature.

Since the Charpy energy C_v cannot be used in a predictive mode, design codes use the energy and the transition temperature in setting materials' specifications for applications. For example, the ASME Boiler and Pressure Vessel Code Section VIII, Division 3

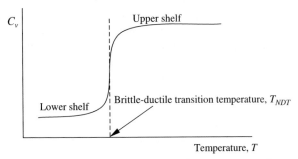

Figure 1.3 Variation of the Charpy V-notch energy with temperature for typical low-strength ferrous alloys. An abrupt transition to a high-energy ductile fracture mode appears at temperatures above the brittle–ductile transition temperature.

specifies the minimum required Charpy energy C_v at the minimum design metal temperature (MDMT). Empirical correlations of the Charpy V-notch energy to plane strain fracture toughness have been proposed in the brittle fracture region of the Charpy test. For example, the ASME Boiler and Pressure Vessel Code provides the following correlation when the upper shelf regime is above MDMT

$$\left(\frac{K_{IC}}{\sigma_Y}\right)^2 = 5\left[\frac{C_v}{\sigma_Y} - 0.05\right] \tag{1.2}$$

where C_v is in ft lbf (1 ft lbf = 1.35582 J), the yield stress, σ_Y, is given in ksi and the plane strain fracture toughness K_{IC} is given in ksi $\sqrt{\text{in}}$ (1 ksi $\sqrt{\text{in}}$ = 6.89475 MPa $\sqrt{\text{m}}$). According to the code, the value of the fracture toughness estimated using Eq. 1.2 can be used in the evaluation of fracture criticality under slow-loading K_{IC} conditions. It should be noted that this approach does not take into account the possible rate dependence of crack initiation, growth and possible arrest of the cracks.

Regardless of the attempts to tie the Charpy test results to quantitative fracture mechanics parameters, this class of fracture parameters remains empirical; the utility of these methods and results lies in correlating with past experience, and in qualitative ranking of different materials, and not in providing a predictive methodology for the evaluation of the integrity of structures containing cracks and loaded dynamically. The latter is clearly in the realm of dynamic fracture mechanics.

The above discussion presents a sampling of the problems in which dynamic fracture plays a key role. These and other practical applications have driven the research into dynamic fracture problems. In this book, fundamental mechanics aspects of dynamic fracture are presented. In the first segment, the basic concepts from the continuum theory of dynamic fracture are presented; in Chapter 2, a review of linear elastodynamics is presented. This is followed in Chapter 3 by a description of the stress field in the vicinity of the crack tip and the idea of the dynamic stress intensity factor is discussed. A concise description of the analytical determination of the dynamic stress intensity factor is provided in Chapter 4. This is followed in Chapter 5 by a consideration of the dynamic energy rate balance criterion for the formulation of dynamic fracture criteria. In the second segment, experimental and practical aspects of dynamic fracture are addressed. Methods of generating well-characterized dynamic loads for investigations of dynamic fracture phenomena are described in Chapter 6. This is followed by a detailed exposition of the methods of measuring crack speeds in Chapters 7 and of the diagnostic tools for characterization of the crack tip stress and or deformation fields in Chapter 8. The applicability of the concept of the dynamic stress intensity factor in characterizing dynamic fracture is discussed in Chapter 9. Experimental investigations aimed at formulating dynamic fracture criteria—separated into criteria for crack initiation, growth and arrest—are discussed in Chapter 10; practical applications of these ideas in assessment of structural integrity are developed as well. In the last segment, the physical aspects and models of dynamic fracture are discussed. Chapter 11 is devoted to a discussion of the mechanisms of fracture in different materials; discrepancies between the continuum theory and its mechanistic resolution are described. Various models developed for incorporation of the effects of the fracture process zone in the simulation of dynamic fracture are discussed in Chapter 12.

Chapter 2

Linear Elastodynamics

2.1 Fundamental Boundary-Initial Value Problems in Elastodynamics

We begin with a brief description of the linear elastodynamic theory. Complete treatments of the topic including solution techniques and details of the classical solutions can be found in the monographs by Graff (1975), Achenbach (1973) and Miklowitz (1978). Consider a body occupying the region R with boundaries ∂R. Let the displacement vector $\mathbf{u}$ depend on the position vector $\mathbf{x}$ and time t and be denoted by $\mathbf{u}(\mathbf{x}, t)$. The strain tensor $\boldsymbol{\varepsilon}(\mathbf{x}, t)$ is the symmetric gradient of $\mathbf{u}$

$$\boldsymbol{\varepsilon}(\mathbf{x}, t) = \frac{1}{2}[\nabla \mathbf{u} + (\nabla \mathbf{u})^{T}] \tag{2.1}$$

Here we consider the infinitesimal strain tensor and therefore neglect higher order terms involving higher powers of the gradient of $\mathbf{u}$. It is also assumed that $\mathbf{u}$ is a continuous function of $\mathbf{x}$, but it is not required that its derivatives with respect to $\mathbf{x}$ and t be continuous as we shall see later. The material of the body is assumed to be homogeneous, isotropic and linearly elastic. Therefore, the stress tensor $\boldsymbol{\sigma}(\mathbf{x}, t)$ is given by[1]

$$\boldsymbol{\sigma}(\mathbf{x}, t) = \lambda \varepsilon_{kk} \mathbf{1} + 2\mu \boldsymbol{\varepsilon} \tag{2.2}$$

where $\mathbf{1}$ is the identity tensor and λ and μ are the Lamé constants. The balance of linear momentum results in the following equation of motion

$$\nabla \cdot \boldsymbol{\sigma} + \mathbf{f} = \rho \ddot{\mathbf{u}} \tag{2.3}$$

where ρ is the mass density, $\mathbf{f}$ the body force per unit volume and the superdot indicates time derivatives. Symmetry of the stress tensor ensures the balance of angular momentum.

Substituting Eq. 2.1 in Eq. 2.2 and the result in Eq. 2.3, the equations of motion can be obtained in terms of the displacements alone; these are the Navier's equations of motion

$$(\lambda + \mu)\nabla(\nabla \cdot \mathbf{u}) + \mu \nabla^{2} \mathbf{u} + \mathbf{f} = \rho \ddot{\mathbf{u}} \tag{2.4}$$

---

[1] Standard index notation will be used throughout this book. Latin subscripts take the range 1,2,3 while Greek subscripts take the range 1,2. Repeated index implies summation over the range of the index and an index following a comma indicates partial differentiation with respect to the coordinate identified by that index.

This is a system of three partial differential equations governing the motion of points in the body. In this book we shall assume that the body forces vanish and remove them from consideration in subsequent equations. To the set of equations 2.4, we must add initial conditions as well as boundary conditions. As in the quasi-static problem, there are three fundamental problems that can be posed, depending on whether the displacements, tractions or some combination are prescribed on the boundaries. For the displacement boundary value problem

$$\mathbf{u}(\mathbf{x}, t) = \mathbf{u}^*(\mathbf{x}, t) \tag{2.5}$$

on ∂R for $t > 0$, where $\mathbf{u}^*(\mathbf{x}, t)$ is a prescribed function. For the traction boundary value problem, the traction vector, $\mathbf{s}(\mathbf{x}, t)$, is prescribed

$$\mathbf{s}(\mathbf{x}, t) = \boldsymbol{\sigma}(\mathbf{x}, t)\mathbf{n} = \mathbf{s}^*(\mathbf{x}, t) \tag{2.6}$$

on ∂R for $t > 0$, where $\mathbf{n}$ is the unit outward normal and $\mathbf{s}^*(\mathbf{x}, t)$ a prescribed function. The third problem is the mixed-boundary value problem for which the displacements are prescribed in a part of the boundary, and tractions are prescribed over the remainder

$$\mathbf{u}(\mathbf{x}, t) = \mathbf{u}^*(\mathbf{x}, t) \text{ on } \partial_1 R$$
$$\mathbf{s}(\mathbf{x}, t) = \mathbf{s}^*(\mathbf{x}, t) \text{ on } \partial_2 R \tag{2.7}$$

For all three problems, initial conditions must be added to complete the formulation of the problems

$$\mathbf{u}(\mathbf{x}, 0) = \mathbf{u}_0(\mathbf{x})$$
$$\dot{\mathbf{u}}(\mathbf{x}, 0) = \dot{\mathbf{u}}_0(\mathbf{x}) \tag{2.8}$$

on R, where $\mathbf{u}_0(\mathbf{x})$ and $\dot{\mathbf{u}}_0(\mathbf{x})$ are prescribed functions. It should be evident that even though we have written the governing equations in terms of displacements, the boundary conditions may be in terms of tractions, i.e. in terms of linear combinations of the derivatives of the displacement components and therefore complicating the solution of the problem.

Finally, the principle of conservation of energy (or the theorem of power expended) may be written as

$$\int_{\partial R} \mathbf{s} \cdot \frac{\partial \mathbf{u}}{\partial t} \, dR + \int_R \rho \mathbf{f} \cdot \frac{\partial \mathbf{u}}{\partial t} \, dV = \frac{d}{dt} [U(t) + T(t)] \tag{2.9}$$

where $U(t)$ is the strain energy and $T(t)$ the kinetic energy given by

$$U(t) = \int_R \frac{1}{2} \boldsymbol{\sigma} \cdot \boldsymbol{\varepsilon} \, dV \tag{2.10}$$

$$T(t) = \int_R \frac{1}{2} \rho \dot{\mathbf{u}} \cdot \dot{\mathbf{u}} \, dV \tag{2.11}$$

For problems in classical elastodynamics the conservation of energy provides a convenient way of approaching solutions; however, since there are no dissipative processes, it is seldom necessary to introduce the energy conservation equations explicitly into the problem formulation. On the other hand, in fracture problems that are the focus of this book, dissipation is inherent in the problem. The fracture processes that occur in the crack tip region remove energy from the system and hence, we have to augment Eq. 2.9 to account for the dissipation that occurs in the fracture process regions as we shall discuss later. Boundary-initial value problems posed within the context of linear elastodynamics above possess unique solutions (see Wheeler and Sternberg, 1968).

2.2 Bulk Waves

Eq. 2.4 represents a hyperbolic system of partial differential equations and hence admit propagating wave solutions. The character of these waves can be obtained by considering special deformations. The Laplacian of $\mathbf{u}$ in Eq. 2.4 can be replaced using the following vector identity $\nabla^2\mathbf{u} = \nabla(\nabla\cdot\mathbf{u}) - \nabla\times\nabla\times\mathbf{u}$ to yield

$$(\lambda + 2\mu)\nabla(\nabla\cdot\mathbf{u}) - \mu\nabla\times\nabla\times\mathbf{u} = \rho\ddot{\mathbf{u}} \qquad (2.12)$$

First, if we consider $\mathbf{u}$ to be an irrotational deformation, $\nabla\times\mathbf{u} = 0$, Eq. 2.4 reduces to

$$\nabla^2\mathbf{u} = \frac{1}{C_d^2}\ddot{\mathbf{u}} \text{ with } C_d = \sqrt{\frac{\lambda + 2\mu}{\rho}} \qquad (2.13)$$

This deformation is seen to obey the standard wave equation with a characteristic speed C_d. Such waves are called *irrotational* or *dilatational waves*. Next, if we consider the dilatation to be zero, $\nabla\cdot\mathbf{u} = 0$, we obtain the case of an equivoluminal deformation. In this case, Eq. 2.4 becomes

$$\nabla^2\mathbf{u} = \frac{1}{C_s^2}\ddot{\mathbf{u}} \text{ with } C_s = \sqrt{\frac{\mu}{\rho}} \qquad (2.14)$$

Therefore, equivoluminal deformations also obey the standard wave equation but with a characteristic speed C_s. Such waves are called *equivoluminal* or *shear waves*. Clearly $C_d > C_s$; an observer at some distance from a source (such as an earthquake) of these waves will first receive the dilatational wave and then the equivoluminal wave; hence in the seismology literature these waves are called *primary (P)* waves and *secondary (S)* waves. While the wave speeds are expressed here in terms of the Lamé constants, materials are usually characterized in terms of the engineering constants E and ν, the modulus of elasticity and Poisson's ratio, respectively. Conversion between these constants can be effected using the following relationships

$$E = \frac{\mu(3\lambda + 2\mu)}{\lambda + \mu}, \nu = \frac{\lambda}{2(\lambda + \mu)} \qquad (2.15)$$

Now, the wave speeds may be expressed in terms of E and ν, but more importantly, the ratio of wave speeds is seen to depend only on the Poisson's ratio

$$\frac{C_d}{C_s} = \left(\frac{2 - 2\nu}{1 - 2\nu}\right)^{1/2} \equiv k \qquad (2.16)$$

Representative values of material properties are given in Table 2.1; these values are based on nominal values of modulus of elasticity, Poisson's ratio and density in order to provide

Table 2.1 Wave speeds in solids

Material	Modulus of elasticity, E (GPa)	Poisson's ratio, ν	Density, ρ (Mg/m^3)	Dilatational wave speed, C_d (m/s)	Distortional wave speed, C_s (m/s)	Plane stress dilatational wave speed, C_d^p (m/s)	Rayleigh wave speed, C_R (m/s)
High strength steel	200	0.3	7.8	5875	3140	5308	2913
Tungsten and alloys	406	0.3	13.4	6386	3414	5770	3167
Aluminum alloys	70	0.3	2.7	5908	3158	5338	2929
Alumina	390	0.22	3.9	10,685	6402	10,251	5858
Silicon nitride	350	0.22	3.2	11,175	6695	10,721	6127
Silica glass	70	0.22	2.6	5544	3322	5319	3040
Homalite-100	4.5	0.34	1.2	2402	1183	2059	1104
Plexiglas	3.4	0.34	1.2	2088	1028	1790	960
Polycarbonate	2.6	0.40	1.2	1826	899	1565	839
Rubber	0.1	0.499	0.85	4434	198	396	189
	0.01	0.499	0.85	1402	63	125	60

an idea about the order of magnitude of the wave speeds. As can be seen from the values in the table, the shear wave speeds are typically about one-half of the dilatational wave speeds.

2.3 Lamé Solution

Since the problem considered here is linear, Eqs. 2.13 and 2.14 suggest that propagation of an arbitrary deformation that is a combination of dilatation and shear will be governed by both types of waves. This can be shown directly using the Lamé solution of the displacement equations of motion (Eq. 2.4). Consider the following representation of the displacement vector **u**

$$\mathbf{u} = \nabla\varphi + \nabla \times \boldsymbol{\psi} \tag{2.17}$$

where $\varphi(\mathbf{x}, t)$ is a scalar function and $\boldsymbol{\psi}(\mathbf{x}, t)$ a vector-valued function with $\nabla \cdot \boldsymbol{\psi} = 0$. The displacement components obtained from Eq. 2.17 will satisfy the differential equations (Eq. 2.4) if $\varphi(\mathbf{x}, t)$ and $\boldsymbol{\psi}(\mathbf{x}, t)$ are obtained as solutions of the following wave equations

$$\nabla^2\varphi = \frac{1}{C_d^2}\ddot{\varphi} \tag{2.18}$$

$$\nabla^2\boldsymbol{\psi} = \frac{1}{C_s^2}\ddot{\boldsymbol{\psi}} \tag{2.19}$$

The scalar potential $\varphi(\mathbf{x}, t)$ corresponds to the dilatational wave and the vector potential $\boldsymbol{\psi}(\mathbf{x}, t)$ corresponds to shear waves. The completeness of the Lamé decomposition of the displacement vector has been demonstrated by Clebsch, Somigliana and others. Sternberg (1960) provides a discussion of this decomposition.

It should be noted that the Lamé solution is really a reduction of the complicated hyperbolic system of equations for the displacement vector into two standard wave equations for the potentials $\varphi(\mathbf{x}, t)$ and $\boldsymbol{\psi}(\mathbf{x}, t)$ coupled through the boundary conditions. As Miklowitz points out, the advantage of reformulating Navier's equations in terms of the Lamé potentials is that solutions and solution procedures developed for the standard wave equation can now be used to address problems associated with elastic wave propagation in solids.

2.4 Plane Waves

In order to gain insight into the wave character of the dynamic problem, consider the propagation of plane waves in a three-dimensional solid medium. A plane is defined by $\mathbf{x} \cdot \mathbf{n} = d$ where $\mathbf{x}$ represents the position vector of any point in the plane, $\mathbf{n}$ the normal to the plane and d the distance from the origin to the plane along the normal. If the plane is assumed to move in the direction of the normal at a wave speed c, then $d = d_0 + ct$, where d_0 is the location of the plane at time $t = 0$, describes the propagation of the plane. Clearly, as the wave propagates, d_0 remains constant and is called the *phase*; the surface with constant phase (in this case the plane) is the *wavefront*. Now, applying this idea to elastodynamics, plane waves corresponding to dilatational and shear deformations can be represented as

$$\varphi = \varphi(\mathbf{x} \cdot \mathbf{n} - C_d t) \tag{2.20}$$

$$\boldsymbol{\psi} = \boldsymbol{\psi}(\mathbf{x} \cdot \mathbf{n} - C_s t) \tag{2.21}$$

It is easily demonstrated by substitution that if $\varphi(\mathbf{x}, t)$ and $\boldsymbol{\psi}(\mathbf{x}, t)$ are represented as above, they automatically satisfy the wave equations 2.18 and 2.19, respectively. To examine the plane waves further, without loss of generality, the direction of propagation can be taken to be the x_1 axis; then using Eqs. 2.20 and 2.21 in Eq. 2.17, we obtain the displacement components

$$
\begin{aligned}
u_1 &= \varphi'(x_1 - C_d t) \\
u_2 &= -\psi_3'(x_1 - C_s t) \\
u_3 &= \psi_2'(x_1 - C_s t)
\end{aligned}
\tag{2.22}
$$

where the prime denotes differentiation with respect to the argument. The dilatational wave travels at the speed C_d and can only sustain a displacement u_1 in the direction of wave propagation, x_1; hence this is a longitudinal wave with the particle motion in the direction of the wave propagation. This is the P wave or dilatational; note that it could be a compressive or tensile wave depending on whether the particle motion is in the direction of motion or opposed to it. The shear wave travels at the speed C_s and can sustain displacement components u_2 and u_3, i.e. in the directions perpendicular to the wave propagation; hence these are transverse waves. As a consequence of our resolution of the vector along the Cartesian coordinates, the shear wave has been decomposed into two components, with particle motions in x_2 and x_3. These are commonly called the SH arid SV waves (for the *horizontally polarized* and *vertically polarized* shear waves where the x_3 axis points in the vertical direction).

2.5 Propagation of Discontinuities: Wavefronts and Rays

The displacement vector $\mathbf{u}(\mathbf{x}, t)$ need not possess continuous derivatives; the governing equations 2.4 allow discontinuities in the derivatives of $\mathbf{u}(\mathbf{x}, t)$ to exist along certain planes (called *wavefronts*) and propagate along certain directions (called *rays*). Discontinuities in the spatial gradients of $\mathbf{u}(\mathbf{x}, t)$ imply a discontinuity in the strains and stresses, and discontinuities in the temporal gradients of $\mathbf{u}(\mathbf{x}, t)$ indicate jumps in the particle velocity and/or acceleration. While such discontinuities cannot be sustained physically, rapid changes in $\mathbf{u}(\mathbf{x}, t)$ that occur over very short distances or time intervals are approximated as discontinuous jumps in the gradients. This representation is useful in characterizing the variations in the strains, stresses and velocities generated by suddenly applied loads. Love (1927) described the kinematic and dynamic conditions that must hold on a surface of discontinuity. With the normal to the discontinuity denoted by $\mathbf{n}$, the kinematic and dynamic jump conditions are

$$[\dot{u}_i] = -c[u_{i,j}]n_j \tag{2.23}$$

$$-\rho c[\dot{u}_i] = [\sigma_{ij}]n_j \tag{2.24}$$

where ρ is the density, c the appropriate wave speed and the square bracket around a quantity indicates the jump in that quantity across the discontinuity. Eqs. 2.23 and 2.24 can be interpreted using the equations of motion 2.4. Introducing Eq. 2.4 in Eq. 2.24 yields

$$\rho c[\dot{u}_i] = -\lambda\delta_{ij}[u_{k,k}]n_j - \mu[u_{i,j}]n_j - \mu[u_{j,i}]n_j \tag{2.25}$$

which may be rearranged as follows:

$$(\rho c^2 - \mu)[\dot{u}_i] = -c(\lambda + \mu)[u_{k,k}]n_j \tag{2.26}$$

If we impose a velocity jump $[\dot{u}_i]$ with zero dilatation, $[u_{k,k}] = 0$, the jump propagates at a speed $c = C_s$. In a similar manner, if we consider a velocity jump $[\dot{u}_i]$ with $\nabla \times \mathbf{u} = 0$, Eq. 2.25 reduces to

$$(\rho c^2 - (\lambda + 2\mu))[\dot{u}_i]n_i = 0 \tag{2.27}$$

which indicates that jumps in dilatation travel with the speed $c = C_d$. Therefore, we might expect that if an arbitrary velocity jump is provided (through an external loading agent or from an internal source), both dilatational and shear waves propagate in the body carrying the appropriate jump discontinuities along both wavefronts.

The construction of wavefronts and rays is useful in understanding and interpreting the development of stress fields in elastodynamic problems. So, we shall briefly outline the construction of the equations for the wavefronts and rays. The surface of discontinuity may be written as $S(\mathbf{x}, t) = \tau(\mathbf{x}) - t = 0$ or equivalently by $t = \tau(\mathbf{x})$. At any point on the wave front, the ray is normal to the wavefront; thus, the governing equation for the rays is obtained

$$\frac{d\mathbf{x}}{dt} = c\mathbf{n} = c\frac{\nabla\tau(\mathbf{x})}{|\nabla\tau(\mathbf{x})|} \tag{2.28}$$

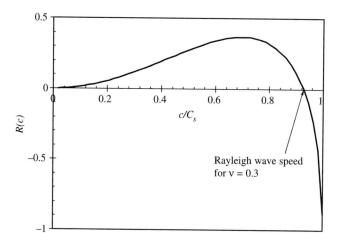

Figure 2.1 Variation of the Rayleigh function $R(c)$ with speed.

This is a cubic equation for k_R^2, and depends only on the Poisson's ratio. The roots of this equation indicate the propagation speed of the surface wave assumed in Eq. 2.46. Physically meaningful solutions to Eq. 2.51 must be in the range $0 < k_R < 1$. For Poisson's ratio in the range of $0 < \nu < 0.5$, at least one real solution exists. This solution corresponds to the Rayleigh surface wave. Viktorov (1967) developed an approximate representation for the Rayleigh wave speed, C_R

$$k_R = \frac{C_R}{C_s} = \frac{0.862 + 1.14\nu}{1 + \nu} \tag{2.53}$$

Clearly, $C_R < C_s$. Table 2.1 lists the values of Rayleigh wave speed for selected materials. Eqs. 2.48 and 2.49 indicate that the Rayleigh wave travels along the surface in the x_1 direction, and experiences an exponential decay along the x_2 direction.

In this chapter, we have presented a formal statement for the fundamental problems and described the nature of propagating waves within this theory; solving boundary value problems under dynamical loading still requires enormous effort. Typically integral transform methods and Green's function methods are used in obtaining solutions to specific boundary value problems. For propagating cracks, further complication of moving boundary conditions must be addressed. In the following sections, we shall describe some of the methods used and solutions obtained for a few crack problems.

2.8 Half-Space Green's Functions

We begin by deriving the half-space Green's function for elastodynamics and then discuss the approaches for solving dynamic crack growth problems. The region of interest is $x_2 \geq 0$. The governing differential equations for the potentials $\varphi = \varphi(x_1, x_2, t)$ and

$\psi(x_1, x_2, t)$ are the wave equations in Eqs. 2.38 and 2.39. Transform techniques are employed for obtaining solutions to these equations. The Laplace transform of a function $f(x_1, x_2, t)$ is defined as

$$\hat{f}(x_1, x_2, s) = \int_0^\infty f(x_1, x_2, t) e^{-st} dt \tag{2.54}$$

where s is the transform parameter, considered to be real and positive. The bilateral Laplace transform is defined as

$$F(\zeta, x_2, s) = \int_{-\infty}^\infty \hat{f}(x_1, x_2, s) e^{-s\zeta x_1} dx_1 \tag{2.55}$$

where $s\zeta$ is the transform parameter and ζ is complex. There are variations in the definition of the bilateral Laplace transform; Kostrov, e.g. uses a parameter q instead of $s\zeta$; the use of $s\zeta$ is a matter of convenience since by proper choice of notation s can be eliminated from the equations in some problems as can be observed in the following development. Applying the Laplace transforms indicated in Eq. 2.54 the wave equations are transformed into the following

$$\nabla^2 \hat{\varphi} - a^2 s^2 \hat{\varphi} = 0, \ \nabla^2 \hat{\psi} - b^2 s^2 \hat{\psi} = 0 \tag{2.56}$$

where $a = 1/C_d$ and $b = 1/C_s$ are the longitudinal and shear wave slownesses, respectively. Next, the bilateral Laplace transform defined in Eq. 2.55 is applied to yield two ordinary differential equations

$$\frac{d^2 \Phi}{dx_2^2} - \alpha^2(\zeta) s^2 \Phi = 0, \ \frac{d^2 \Psi}{dx_2^2} - \beta^2(\zeta) s^2 \Psi = 0 \tag{2.57}$$

where $\Phi(\zeta, x_2, s)$ and $\Psi(\zeta, x_2, s)$ are the transformed potentials and

$$\alpha(\zeta) = \sqrt{a^2 - \zeta^2}, \ \beta(\zeta) = \sqrt{b^2 - \zeta^2} \tag{2.58}$$

In the following, we suppress the arguments of $\alpha(\zeta)$ and $\beta(\zeta)$ and simply write α and β. Taking the Laplace transform in time and the bilateral Laplace transform in space, the stress-potential and displacement-potential relations in Eqs. 2.40 and 2.41 yield

$$
\begin{aligned}
U_1(\zeta, x_2, s) &= \left[s\zeta\Phi + \frac{d\Psi}{dx_2} \right] \\
U_2(\zeta, x_2, s) &= \left[\frac{d\Phi}{dx_2} - s\zeta\Psi \right] \\
\Sigma_{22}(\zeta, x_2, s) &= \mu \left[\left(\frac{b^2}{a^2} - 2 \right) s^2 \zeta^2 \Phi + \frac{b^2}{a^2} \frac{d^2\Phi}{dx_2^2} - 2s\zeta \frac{d\Psi}{dx_2} \right] \\
\Sigma_{12}(\zeta, x_2, s) &= \mu \left[2s\zeta \frac{d\Phi}{dx_2} + \frac{d^2\Psi}{dx_2^2} - s^2 \zeta^2 \Psi \right]
\end{aligned}
\tag{2.59}
$$

The solutions to the ordinary differential equations in Eq. 2.57 are

$$\Phi(\zeta, x_2, s) = A(\zeta, s)e^{-s\alpha x_2} + A_1(\zeta, s)e^{s\alpha x_2}$$
$$\Psi(\zeta, x_2, s) = B(\zeta, s)e^{-s\beta x_2} + B_1(\zeta, s)e^{s\beta x_2}$$

(2.60)

The exponentially growing terms will be unbounded as $x_2 \to \infty$ and must therefore be rejected. The next task is to determine $A(\zeta, s)$ and $B(\zeta, s)$ by imposing the transformed boundary conditions. Let us consider that the tractions on the surface of the half-space are given by

$$\sigma_{12}(x_1, 0^+, t) = \sigma_1(x_1, t),$$
$$\qquad \qquad \qquad \qquad \qquad \text{for} -\infty < x_1 < \infty$$
$$\sigma_{22}(x_1, 0^+, t) = \sigma_2(x_1, t)$$

(2.61)

where $\sigma_1(x_1, t)$ and $\sigma_2(x_1, t)$ are prescribed functions. Note that we are considering a half-space problem and not crack problem; thus the traction components are specified all along the half-space boundary. Taking the Laplace transform in time and the bilateral Laplace transform in space of the specified boundary conditions results in

$$\Sigma_{12}(\zeta, 0^+, s) = \Sigma_1(\zeta, s)$$
$$\Sigma_{22}(\zeta, 0^+, s) = \Sigma_2(\zeta, s)$$

(2.62)

Substituting from Eqs. 2.62 and 2.60 in Eq. 2.59 results in the following two equations for the unknown functions $A(\zeta, s)$ and $B(\zeta, s)$

$$s^2\mu[(b^2 - 2\zeta^2)A(\zeta, s) + 2\zeta\beta B(\zeta, s)] = \Sigma_2(\zeta, s)$$
$$s^2\mu[-2\zeta\alpha A(\zeta, s) + (b^2 - 2\zeta^2)B(\zeta, s)] = \Sigma_1(\zeta, s)$$

(2.63)

Solving for the unknowns yields

$$A(\zeta, s) = \frac{1}{s^2\mu R(\zeta)}[(b^2 - 2\zeta^2)\Sigma_2(\zeta, s) - 2\zeta\beta\Sigma_1(\zeta, s)]$$
$$B(\zeta, s) = \frac{1}{s^2\mu R(\zeta)}[(b^2 - 2\zeta^2)\Sigma_1(\zeta, s) + 2\zeta\alpha\Sigma_2(\zeta, s)]$$

(2.64)

where $R(\zeta) = (b^2 - 2\zeta^2)^2 + 4\zeta^2\alpha\beta$ is the Rayleigh function already encountered in Eq. 2.51. With $A(\zeta, s)$ and $B(\zeta, s)$ given above, it is now possible to write formally the general expressions for the stress and displacement field components in $x_2 > 0$ in terms of the applied tractions on $x_2 = 0$. Substituting Eq. 2.64 in Eq. 2.60 and then into Eq. 2.59 results in the following expressions for the displacement components

$$U_1(\zeta, x_2, s) = \frac{1}{\mu s R(\zeta)}[\zeta e^{-s\alpha x_2}\{(b^2 - 2\zeta^2)\Sigma_2(\zeta, s)\text{sgn}(x_2) - 2\zeta\beta\Sigma_1(\zeta, s)\}$$
$$\qquad\qquad - \beta e^{-s\beta x_2}\{(b^2 - 2\zeta^2)\Sigma_1(\zeta, s) + 2\zeta\alpha\Sigma_2(\zeta, s)\text{sgn}(x_2)\}]$$

$$U_2(\zeta, x_2, s) = -\frac{1}{\mu s R(\zeta)}[\alpha e^{-s\alpha x_2}\{(b^2 - 2\zeta^2)\Sigma_2(\zeta, s) - 2\zeta^2\beta\Sigma_1(\zeta, s)\text{sgn}(x_2)\}$$
$$\qquad\qquad + \zeta e^{-s\beta x_2}\{(b^2 - 2\zeta^2)\Sigma_1(\zeta, s)\text{sgn}(x_2) + 2\zeta\alpha\Sigma_2(\zeta, s)\}]$$

(2.65)

Eq. 2.65 provides the transform of the displacement fields in terms of the transforms of the applied tractions on $x_2 = 0$. Introduction of the sign function $\mathrm{sgn}(x_2)$ in Eqs. 2.65 allows the above formulas to be used for both $x_2 > 0$ and $x_2 < 0$. Similar expressions may be written for the stress components as well. Inversion of these transforms poses significant challenges and can be accomplished only in some special cases.

We specialize the above equations for $x_2 = 0$, with the anticipation that these are needed for solving crack problems; the displacement on the surface $x_2 = 0$ may then be written as

$$U_1(\zeta, 0, s) = \Gamma_{11}(\zeta, s)\Sigma_1(\zeta, s) + \Gamma_{12}(\zeta, s)\Sigma_2(\zeta, s) \tag{2.66}$$

$$U_2(\zeta, 0, s) = \Gamma_{21}(\zeta, s)\Sigma_1(\zeta, s) + \Gamma_{22}(\zeta, s)\Sigma_2(\zeta, s) \tag{2.67}$$

where:

$$\Gamma_{11}(\zeta, s) = -\frac{b^2\sqrt{b^2 - \zeta^2}}{\mu s R(\zeta)} \tag{2.68}$$

$$\Gamma_{22}(\zeta, s) = -\frac{b^2\sqrt{a^2 - \zeta^2}}{\mu s R(\zeta)} \tag{2.69}$$

$$\Gamma_{12}(\zeta, s) = -\Gamma_{21}(\zeta, s) = \frac{\zeta(b^2 - 2\zeta^2 - 2\alpha\beta)}{\mu s R(\zeta)} \tag{2.70}$$

$\Gamma_{\alpha\beta}(\zeta, s)$ are the transforms of the half-space Green's functions $G_{\alpha\beta}(x_1, t)$. In Eqs. 2.66 and 2.67, the transforms of the displacements are given as the product of two transforms; therefore the displacements must be the double convolution of the original functions

$$u_1(x_1, t) = \int_{-\infty}^{\infty}\int_0^{\infty}[G_{11}(\xi - x_1, t - \tau)\sigma_1(\xi, \tau) + G_{12}(\xi - x_1, t - \tau)\sigma_2(\xi, \tau)]d\tau d\xi$$
$$\equiv G_{11} * \sigma_1 + G_{12} * \sigma_2 \tag{2.71}$$

$$u_2(x_1, t) = \int_{-\infty}^{\infty}\int_0^{\infty}[G_{21}(\xi - x_1, t - \tau)\sigma_1(\xi, \tau) + G_{22}(\xi - x_1, t - \tau)\sigma_2(\xi, \tau)]d\tau d\xi$$
$$\equiv G_{21} * \sigma_1 + G_{22} * \sigma_2 \tag{2.72}$$

where the $*$ stands for the double convolution integral. Thus, for any traction boundary value problem, since $\sigma_1(x_1, t)$ and $\sigma_2(x_1, t)$ are prescribed on $x_2 = 0$, $-\infty < x_1 < \infty$, $t > 0$, the displacements on $x_2 = 0$ can be found from the convolution integrals in Eqs. 2.71 and 2.72; in fact, similar expressions can be obtained for the displacement components in the interior of the body. Thus, the task is to find the inverse transforms of $\Gamma_{\alpha\beta}(\zeta, s)$ corresponding to a unit impulse applied at the origin. This is one of the problems considered by Lamb and the solution is well known. The complete solution of the problem obtained using integral transform methods can be found in the books by Achenbach (1973) and Miklowitz (1978). Slepyan (2002) describes an alternate method for inversion of the Green's functions $\Gamma_{\alpha\beta}(\zeta, s)$. Here we look only at the displacement on the free surface for the Lamb problem.

2.9 Lamb's Problem

Consider a half-plane $x_2 \geq 0$ with a point force applied at the origin as follows

$$\sigma_{22}(x_1, 0, t) = -\delta(x_1)\delta(t),$$
$$\text{for} -\infty < x_1 < \infty \tag{2.73}$$
$$\sigma_{12}(x_1, 0, t) = 0$$

Applying the Laplace transforms to the above boundary conditions and substituting in Eqs. 2.66 and 2.67 results in the following equations for the transforms of the displacement components:

$$U_1(\zeta, s) = \frac{\zeta[b^2 - 2\zeta^2 - 2\zeta\alpha\beta]}{\mu s R(\zeta)}$$
$$\tag{2.74}$$
$$U_2(\zeta, s) = \frac{b^2 \alpha(\zeta)}{\mu s R(\zeta)}$$

Here we consider only the u_2 component of displacement. Formally, the inverse of the bilateral Laplace transform is written as

$$\hat{u}_2(x_1, s) = \frac{1}{2\pi i} \int_{\xi - i\infty}^{\xi + i\infty} \frac{b^2 \alpha(\zeta)}{\mu R(\zeta)} e^{s\zeta x_1} d\zeta \tag{2.75}$$

where $-a < \xi < 0$. Evaluation of this integral is accomplished by invoking the Cagniard-de Hoop technique. The path of the integration is shown in Fig. 2.2. By completing the contour as indicated in the figure, the integral in Eq. 2.75 can be converted to an integral along the real axis. The integrand is analytic inside the closed contour and therefore by Cauchy's theorem, the integral is zero. However, by the decay of the integrand as $\zeta \to \infty$, it is evident that the integral along the circular arcs is zero. Therefore, Eq. 2.75 can be

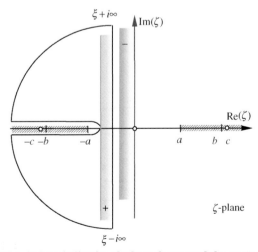

Figure 2.2 Complex ζ plane indicating the branch cuts and the contour of integration.

rewritten as an integral along the real axis

$$\hat{u}_2(x_1, s) = -\frac{b^2}{\pi\mu} \int_{-a}^{-\infty} \text{Im}\left\{\frac{\alpha(\zeta)}{R(\zeta)}\right\} e^{s\zeta x_1} d\zeta \tag{2.76}$$

The Laplace transform in time is inverted by rearranging the integral in such a manner that it represents the product of two Laplace transforms; the inversion is then immediate by the convolution theorem. This is accomplished by noting that Eq. 2.76 can be recast in the form of a Laplace transform if ζx_1 is redefined as $-\eta$

$$\hat{u}_2(x_1, s) = -\frac{b^2}{\pi\mu x_1} \int_{ax_1}^{t} \text{Im}\left\{\frac{\alpha(-\eta/x_1)}{R(-\eta/x_1)}\right\} e^{-s\eta} d\eta \tag{2.77}$$

Evaluation of the above integral is straightforward, but attention should be paid to the time of arrival of the dilatational and distortional waves

$$u_2(x_1, t) = -\frac{b^2}{\pi\mu x_1} \text{Im}\left\{\frac{\alpha(-t/x_1)}{R(-t/x_1)}\right\} \equiv G_{22}(x_1, t) \tag{2.78}$$

From this equation, we obtain

$$G_{22}(x_1, t) = \begin{cases} \dfrac{b^2(b^2 - 2\xi^2)^2\sqrt{\xi^2 - a^2}}{\pi\mu x_1[(b^2 - 2\xi^2)^4 - 16\xi^4\alpha^2\beta^2]} & \text{for } |ax_1| < t < |bx_1| \\[4mm] \dfrac{b^2\sqrt{\xi^2 - a^2}}{\pi\mu x_1 R(\xi)} & t > |bx_1| \end{cases} \tag{2.79}$$

where $\xi = t/x_1$. $G_{22}(\pi\mu x_1/C_s)$ is plotted in Fig. 2.3 as a function of the normalized time $\tau = C_d t/x_1$; this is the normal displacement felt by an observer located at x_1. For $\tau < 1$, the dilatational wave has not reached the observer and hence the displacement is zero.

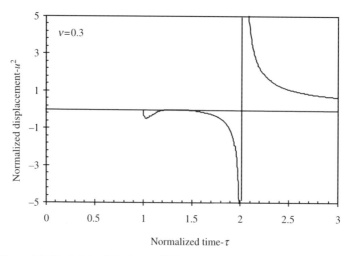

Figure 2.3 Variation of $G_{22}(\pi\mu x_1/C_s)$ with normalized time $\tau = C_d t/x_1$.

The dilatational wave brings a very small upward displacement of the surface; this is followed later by the distortional wave that arrives at $\tau = k$ and closely behind by the Rayleigh wave at $\tau = k/k_R$; the singularity of $R(C_R)$ at this time is seen in Fig. 2.3. Other Green's functions may be constructed in a similar manner; Slepyan (2002) has presented the complete set of expressions for the Green's functions appropriate for the half-space problem considered here. With the Green's function known, any traction distribution on the half-space can be imposed and introduced into Eqs. 2.71 and 2.72 to obtain the surface displacements.

Chapter 3

Dynamic Crack Tip Fields

3.1 Dynamically Loaded Cracks

Within the framework of the two-dimensional linear elastodynamic theory described in Chapter 2, we first formulate general crack problems. Consider an unbounded linearly elastic medium containing a crack. Without loss of generality, we may consider a state of plane strain[1] and assume that the crack lies *initially* along $x_1 < 0$, $x_2 = 0$. As a consequence of the applied loading the crack may extend, but at first we suppress the crack extension. Furthermore, we may assume that the far boundaries of the specimen are traction free and that tractions are applied only on the crack surfaces. Let the crack face boundary conditions be given in terms of the stress components as

$$\sigma_{12}(x_1, 0^\pm, t) = -sH(t)$$
$$\sigma_{22}(x_1, 0^\pm, t) = -pH(t) \quad \text{for } x_1 \le 0 \qquad (3.1)$$
$$\sigma_{32}(x_1, 0^\pm, t) = qH(t)$$

for $x_1 \le 0$, where p, s, and q are prescribed and $H(t)$ is the unit step function. These three loads correspond to conventional modes I, II and III of linear elastic fracture mechanics and are illustrated in Fig. 3.1. Determination of the stress and deformation fields near the crack tip requires the solution of the governing equations for the potentials $\varphi = \varphi(x_1, x_2, t)$ and $\psi(x_1, x_2, t)$ given by the wave equations in Eqs. 2.38 and 2.39.

By the application of Huygens' principle the wavefronts emanating from the crack under this loading can be drawn and are shown in Fig. 3.2. Bulk dilatational and shear waves travel into the body from the crack surface; these wavefronts are parallel to the crack line at distances far from the crack tip. We note that these wavefronts simply carry a jump in the appropriate stress components. On the other hand, the crack tip appears more like a point source and radiates cylindrical dilatational and shear wavefronts in this two-dimensional problem. The headwave or von Schmidt wave generated from the compression wave interacting with the free surface is shown as the angled lines.

---

[1] From the discussion in Chapter 2, it should be clear that the solution corresponding to the plane stress problem can be obtained by replacing the bulk wave speed with the plate wave speed.

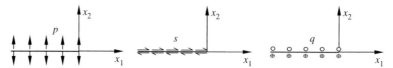

Figure 3.1 Semi-infinite crack under uniform loading representing opening, in-plane shearing and anti-plane shearing loading conditions.

The Rayleigh surface wave also travels along the negative x_1 direction just behind the cylindrical shear wavefront; this wave is indicated in the figure by the small dot. In order to determine the crack tip stresses the elastodynamic field behind the cylindrical wavefronts must be determined.

An alternate way of thinking about the wave loading is to consider the loads along the crack line as a distribution of point sources; a point source at $x_1 = -\xi$ radiates a cylindrical wavefront that reaches the crack tip at $t = \xi/C_d$. Since the load is distributed uniformly over the entire crack line, sources at farther distances from the crack tip will influence the crack tip stress field at later times continuously increasing the stress at the crack tip.

For a stationary crack, the elastodynamic stress field near a crack tip has the same structure as the quasi-static crack tip with the only difference being the time dependence of the stress intensity factor. Thus,

$$\sigma_{\alpha\beta}(r, \theta) = \frac{K_{\mathrm{I}}(t)}{\sqrt{2\pi r}} f^{\mathrm{Is}}_{\alpha\beta}(\theta) + \frac{K_{\mathrm{II}}(t)}{\sqrt{2\pi r}} f^{\mathrm{IIs}}_{\alpha\beta}(\theta) + \cdots \quad \text{as } r \to 0$$

$$\sigma_{3\alpha}(r, \theta) = \frac{K_{\mathrm{III}}(t)}{\sqrt{2\pi r}} f^{\mathrm{IIIs}}_{3\alpha}(\theta)$$

(3.2)

where $K_{\mathrm{I}}(t)$, $K_{\mathrm{II}}(t)$ and $K_{\mathrm{III}}(t)$ are the *dynamic stress intensity factors* under modes I, II and III, respectively; they depend on time as well as the applied loading and geometry and must be determined through a solution to the appropriate initial-boundary value problem.

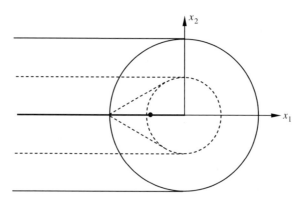

Figure 3.2 Waves emanating from a crack tip. The solid lines indicate the dilatational waves. The dotted lines indicate shear waves.

$f_{\alpha\beta}^{\text{Is}}(\theta), f_{\alpha\beta}^{\text{IIs}}(\theta)$ and $f_{\alpha\beta}^{\text{IIIs}}(\theta)$ are the angular variation of the crack tip stress field and are given by

$$
\begin{aligned}
f_{11}^{\text{Is}}(\theta) &= \cos\tfrac{1}{2}\theta[1 - \sin\tfrac{1}{2}\theta\sin\tfrac{3}{2}\theta] \\
f_{22}^{\text{Is}}(\theta) &= \cos\tfrac{1}{2}\theta[1 + \sin\tfrac{1}{2}\theta\sin\tfrac{3}{2}\theta] \\
f_{12}^{\text{Is}}(\theta) &= \cos\tfrac{1}{2}\theta\sin\tfrac{1}{2}\theta\cos\tfrac{3}{2}\theta
\end{aligned}
\tag{3.3}
$$

$$
\begin{aligned}
f_{11}^{\text{IIs}}(\theta) &= -\sin\tfrac{1}{2}\theta[2 + \cos\tfrac{1}{2}\theta\cos\tfrac{3}{2}\theta] \\
f_{22}^{\text{IIs}}(\theta) &= \cos\tfrac{1}{2}\theta\sin\tfrac{1}{2}\theta\cos\tfrac{3}{2}\theta \\
f_{12}^{\text{IIs}}(\theta) &= \cos\tfrac{1}{2}\theta[1 - \sin\tfrac{1}{2}\theta\sin\tfrac{3}{2}\theta]
\end{aligned}
\tag{3.4}
$$

$$
\begin{aligned}
f_{31}^{\text{IIIs}}(\theta) &= -\sin\tfrac{1}{2}\theta \\
f_{32}^{\text{IIIs}}(\theta) &= \cos\tfrac{1}{2}\theta
\end{aligned}
\tag{3.5}
$$

Since we have suppressed the possibility that the crack may extend, the problem can be solved by standard methods for mixed boundary value problems. As noted earlier, since no dissipative processes are involved, there is no need to invoke the energy balance equation. However, as we have seen above, the crack tip stresses increase continuously with time; these stresses must reach a critical state and the crack must begin to grow dynamically. Once a crack begins to propagate, the direction and speed with which it moves must be determined. Therefore, an equation for the motion of the crack tip must be obtained. Mott (1948) proposed a simple extension of the Griffith criterion: the speed may be determined by including the kinetic energy in the energy balance equation. Writing this equation formally, the crack must extend along a suitable path at a suitable speed in order to obey the energy rate balance equation 2.9 rewritten here as

$$
\int_{\partial R} \mathbf{s}\cdot\frac{\partial \mathbf{u}}{\partial t}\, dR = [U(t) + T(t)] + \frac{dD}{dt}
\tag{3.6}
$$

to include the dissipation at the crack tip. $U(t)$ is the strain energy and $T(t)$ is the kinetic energy defined in Eqs. 2.10 and 2.11. D is the dissipation in the fracture process zone. Eq. 3.6 must provide the criterion for path selection as well as for the speed of the crack along this path. In addition, it must be ensured that the boundary condition in Eq. 3.1 is augmented with traction free conditions on the newly created crack surfaces

$$
\begin{aligned}
\sigma_{22}(x_1, x_2, t) &= 0 \\
\sigma_{12}(x_1, x_2, t) &= 0 \qquad \text{for } (x_1, x_2) \in S \\
\sigma_{23}(x_1, x_2, t) &= 0
\end{aligned}
\tag{3.7}
$$

where S is the unknown extension of the crack. While this is now a physically complete formulation of the dynamic crack problem, it is not quite practical. If a representation for dissipation as described in Eq. 3.6 is readily available or easily developed, numerical solution of the system of equations 2.38 and 2.39 with boundary conditions 3.1 and 3.7 can be considered. Recent developments in the formulation of cohesive zone models for the crack tip dissipative processes have enabled large-scale computational simulations of dynamic fracture problems, but due to the inherent limitations of the cohesive zone models, such simulations have not been able to capture all the observed dynamic fracture phenomena. We shall describe these efforts later.

In this section, we focus our attention on what has been the most successful strategy in dynamic fracture analysis. The fracture criterion (or equivalently the energy rate balance condition in Eq. 3.6) is decoupled from the governing field equations by prescribing the path and speed of the crack tip; typically the crack is assumed to grow along a straight line at a constant speed. While this decoupling makes the problem amenable to analysis, the restriction of rectilinear crack extension is so severe that it is generally appropriate only in situations where the applied load has a mode I symmetry or the crack is trapped by a weak plane in layered media. In this manner, a set of dynamically admissible solutions is obtained; from this set, the correct solution must be selected by imposing the energy balance equation, with suitable assumptions regarding the dissipation to make the problem manageable. We describe the analytical solutions to dynamic problems in two parts—the first part dealing with stationary cracks under dynamic loads and the second part dealing with dynamically growing cracks. Failure criteria will be examined after a description of these analyses.

3.2 Asymptotic Analysis of Crack Tip Fields

The dynamic crack tip field exhibits a square-root singularity just as in the case of the quasi-static problem. In this section, we determine the structure of this singular field for anti-plane shear, opening mode and in-plane shear loading.

3.2.1 Anti-Plane Shear

Consider a traction free crack that is assumed to lie initially along $x_1 < 0$, $x_2 = 0$ and to move along $x_2 = 0$ at a *constant* speed $v < C_R$. The governing differential equation for the nonzero displacement component u_3 is

$$\nabla^2 u_3 = \frac{1}{C_s^2} \ddot{u}_3 \tag{3.8}$$

The traction-free boundary condition can be written as

$$\sigma_{32}(x_1, 0^\pm) = \mu u_{3,2}(x_1, 0^\pm) = 0 \tag{3.9}$$

where $x_2 = 0^\pm$ indicates approach to the crack surface from the positive or negative x_2 direction. If we use a Galilean transformation to a coordinate system moving with the crack tip with $\xi_1 = x_1 - vt$, $\xi_2 = x_2$, Eq. 3.8 becomes

$$\left(1 - \frac{v^2}{C_s^2}\right) \frac{\partial^2 u_3}{\partial \xi_1^2} + \frac{\partial^2 u_3}{\partial \xi_2^2} = 0 \tag{3.10}$$

where $u_3 = u_3(\xi_1, \xi_2)$. Introducing a coordinate scaling $\zeta_s = \xi_1 + i\alpha_s \xi_2$, the governing equation reduces to

$$\nabla^2 u_3(r_s, \theta_s) = 0 \tag{3.11}$$

where

$$r_{\mathrm{s}} = \sqrt{\xi_1^2 + \alpha_{\mathrm{s}}^2 \xi_2^2}, \quad \theta_{\mathrm{s}} = \arctan\left(\frac{\alpha_{\mathrm{s}} \xi_2}{\xi_1}\right), \quad \alpha_{\mathrm{s}} = \sqrt{1 - \frac{v^2}{C_{\mathrm{s}}^2}} \tag{3.12}$$

Let us seek a separable form of the solution for u_3, similar to the Williams (1957) expansion for the quasi-static crack problem

$$u_3(r_{\mathrm{s}}, \theta_{\mathrm{s}}) = r_{\mathrm{s}}^\lambda f(\theta_{\mathrm{s}}; \lambda) \tag{3.13}$$

Eq. 3.11 reduces to an ordinary differential equation for the unknown function f

$$f'' + \lambda^2 f = 0 \tag{3.14}$$

The general solution to Eq. 3.14 that obeys the anti-plane symmetry of the problem is

$$f(\theta_{\mathrm{s}}; \lambda) = A \sin \lambda \theta_{\mathrm{s}} \tag{3.15}$$

where A is a constant. Introducing this solution into the boundary condition in Eq. 3.9 results in the characteristic equation for the determination of λ. As in the quasi-static case, after rejecting the trivial solution, the stress components are singular and the displacements are bounded only when $\lambda = 1/2$. Of course, for $\lambda > 1/2$ both stresses and displacements go to zero as the crack tip is approached; these terms are not considered here, but we shall examine them for the plane strain problem. Thus,

$$u_3(r_{\mathrm{s}}, \theta_{\mathrm{s}}) = 2A r_{\mathrm{s}}^{1/2} \sin\frac{1}{2}\theta_{\mathrm{s}} + \cdots \tag{3.16}$$

The amplitude parameter A is left undetermined in this local analysis and must be obtained from a complete solution of the problem; it can be re-defined in terms of the mode III *dynamic stress intensity factor*, K_{III}

$$K_{\mathrm{III}} = \lim_{\xi_1 \to 0} \sqrt{2\pi\xi_1}\,\sigma_{32}(r_{\mathrm{s}}, 0^\pm) \tag{3.17}$$

The stress components are then written as

$$\sigma_{32}(r_{\mathrm{s}}, \theta_{\mathrm{s}}) = \frac{K_{\mathrm{III}}}{\sqrt{2\pi r}} \frac{1}{\sqrt{\gamma_{\mathrm{s}}}} \cos\frac{1}{2}\theta_{\mathrm{s}}$$

$$\sigma_{31}(r_{\mathrm{s}}, \theta_{\mathrm{s}}) = -\frac{K_{\mathrm{III}}}{\sqrt{2\pi r}} \frac{1}{\alpha_{\mathrm{s}}\sqrt{\gamma_{\mathrm{s}}}} \sin\frac{1}{2}\theta_{\mathrm{s}} \tag{3.18}$$

where

$$\gamma_{\mathrm{s}} = \sqrt{1 - (v \sin\theta/C_{\mathrm{s}})^2} \quad \text{and} \quad \tan\theta_{\mathrm{s}} = \alpha_{\mathrm{s}} \tan\theta \tag{3.19}$$

We note that this is entirely analogous to the quasi-static problem: the stress components exhibit the inverse square root singularity, but the angular distribution is distorted by the speed of the moving crack tip. The stress and displacement fields reduce to the corresponding quasi-static fields in the limit of $v \to 0$.

In the above analysis, it has been assumed that the crack was in steady-state motion; under this condition the dynamic stress intensity factor *must* be constant. Freund (1990) has shown that if the crack moves with a nonuniform speed, the result described above carries over completely, with the only change that the stress intensity factor can now be considered to be a function of time and crack speed, $K_{III}(t, v)$. The effects of the nonuniform motion of the crack appear only in terms of higher order than the singular field considered here.

3.2.2 In-Plane Symmetric Deformation

We now turn to the in-plane problem. As discussed before, plane strain and plane stress problems are indistinguishable except for the difference in the P-wave speed; the asymptotic field is developed for the case of plane strain. The procedure for obtaining the asymptotic stress and displacement fields is identical to the mode III problem described in Section 3.2.1. Again, a traction-free crack is assumed to lie initially along $x_1 < 0$, $x_2 = 0$ and to move along $x_2 = 0$ at a *constant* speed $v < C_R$. The governing differential equations for the potentials $\varphi = \varphi(x_1, x_2, t)$ and $\psi(x_1, x_2, t)$ are the wave equations in Eqs. 2.38 and 2.39. Introducing the Galilean transformation $\xi_1 = x_1 - vt$, $\xi_2 = x_2$, and rescaling the coordinates through $\zeta_d = r_d e^{i\theta_d} = \xi_1 + i\alpha_d\xi_2$, $\zeta_s = r_s e^{i\theta_s} = \xi_1 + i\alpha_s\xi_2$, with

$$r_d = \sqrt{\xi_1^2 + \alpha_d^2\xi_2^2}, \qquad \theta_d = \arctan\left(\frac{\alpha_d\xi_2}{\xi_1}\right), \qquad \alpha_d = \sqrt{1 - \frac{v^2}{C_d^2}} \qquad (3.20)$$

and α_s, r_s and θ_s as defined in Eq. 3.12, the governing equations become

$$\nabla^2\varphi(r_d, \theta_d) = 0, \qquad \nabla^2\psi(r_s, \theta_s) = 0 \qquad (3.21)$$

Once again we seek a separable form of the solution for $\varphi = \varphi(r_d, \theta_d)$ and $\psi = \psi(r_s, \theta_s)$, similar to the Williams expansion for the quasi-static crack problem

$$\varphi(r_d, \theta_d) = r_d^\lambda f(\theta_d; \lambda), \qquad \psi(r_s, \theta_s) = r_s^\lambda g(\theta_s; \lambda) \qquad (3.22)$$

Following the same procedure as in the mode III problem, the solution can be written as

$$\varphi(r_d, \theta_d) = Ar_d^\lambda \cos\lambda\theta_d, \qquad \psi(r_s, \theta_s) = Br_s^\lambda \sin\lambda\theta_s \qquad (3.23)$$

Note that we have only considered solutions that are symmetric with respect to the crack—i.e. a mode I problem. The antisymmetric solution would lead to the crack tip field for a mode II or in-plane shear mode crack; this is described in the next section. Imposing the traction-free boundary condition results in the following equations for the constants A and B

$$(1 + \alpha_s^2)A \cos(\lambda - 2)\pi + 2\alpha_s B \cos(\lambda - 2)\pi = 0$$
$$2\alpha_d A \sin(\lambda - 2)\pi + (1 + \alpha_s^2)B \sin(\lambda - 2)\pi = 0 \qquad (3.24)$$

For nontrivial solutions, the determinant of the above system of equations must be zero; this results in the characteristic equation. The general solution for the characteristic equation is

$$\lambda = \tfrac{1}{2}n + 1, \qquad \text{for } n = 1, 2, 3, \ldots \qquad (3.25)$$

Nonpositive values of n result in displacement singularities at the crack tip and are therefore rejected. In the case of the mode III, we examined only the term $n = 1$. Here, we shall examine all positive values of n in order to develop the higher order terms in the crack tip stress field, since such higher order terms influence experimental schemes commonly used in dynamic fracture investigations. From Eq. 3.24 it can be shown that the constants A and B are related by

$$
B = \begin{cases} -\dfrac{2\alpha_d}{1+\alpha_s^2}A & \text{for } n \text{ odd} \\[2ex] -\dfrac{1+\alpha_s^2}{2\alpha_s}A & \text{for } n \text{ even} \end{cases}
\tag{3.26}
$$

Using Eqs. 3.23, 3.25, and 3.26 in Eqs. 2.40 and 2.41, the displacement components can be determined to be

$$
u_1^n(r,\theta) = A_n\left(1+\frac{n}{2}\right)\left\{r_d^{n/2}\cos\left(\frac{n\theta_d}{2}\right) - \chi_1(n)r_s^{n/2}\cos\left(\frac{n\theta_s}{2}\right)\right\}
$$
$$
u_2^n(r,\theta) = A_n\alpha_d\left(1+\frac{n}{2}\right)\left\{-r_d^{n/2}\sin\left(\frac{n\theta_d}{2}\right) + \chi_2(n)r_s^{n/2}\sin\left(\frac{n\theta_s}{2}\right)\right\}
\tag{3.27}
$$

where

$$
\chi_1(n) = \begin{cases} \dfrac{2\alpha_d\alpha_s}{1+\alpha_s^2} & \text{for } n \text{ odd} \\[2ex] \dfrac{1+\alpha_s^2}{2} & \text{for } n \text{ even} \end{cases}
\quad\text{and}\quad
\chi_2(n) = \begin{cases} \dfrac{2}{1+\alpha_s^2} & \text{for } n \text{ odd} \\[2ex] \dfrac{1+\alpha_s^2}{2\alpha_d\alpha_s} & \text{for } n \text{ even} \end{cases}
\tag{3.28}
$$

The corresponding stress components are

$$
\sigma_{11}^n(r,\theta) = \frac{\mu A_n}{(1+\alpha_s^2)}\frac{n}{2}\left(1+\frac{n}{2}\right)\left\{(1+\alpha_s^2)(1+2\alpha_d^2-\alpha_s^2)r_d^{(n/2)-1}\cos\left(\frac{n-2}{2}\theta_d\right)\right.
$$
$$
\left. -\kappa(n)r_s^{(n/2)-1}\cos\left(\frac{n-2}{2}\theta_s\right)\right\}
$$
$$
\sigma_{22}^n(r,\theta) = \frac{\mu A_n}{(1+\alpha_s^2)}\frac{n}{2}\left(1+\frac{n}{2}\right)\left\{-(1+\alpha_s^2)^2 r_d^{(n/2)-1}\cos\left(\frac{n-2}{2}\theta_d\right)\right.
$$
$$
\left. +\kappa(n)r_s^{(n/2)-1}\cos\left(\frac{n-2}{2}\theta_s\right)\right\}
\tag{3.29}
$$
$$
\sigma_{12}^n(r,\theta) = \frac{\mu A_n}{2\alpha_s}\frac{n}{2}\left(1+\frac{n}{2}\right)\left\{-4\alpha_d\alpha_s r_d^{(n/2)-1}\sin\left(\frac{n-2}{2}\theta_d\right)\right.
$$
$$
\left. +\kappa(n)r_s^{(n/2)-1}\sin\left(\frac{n-2}{2}\theta_s\right)\right\}
$$

where

$$\kappa(n) = \begin{cases} 4\alpha_d\alpha_s & \text{for } n \text{ odd} \\ (1 + \alpha_s^2)^2 & \text{for } n \text{ even} \end{cases} \tag{3.30}$$

Clearly the term corresponding to $n = 1$ results in bounded displacements and inverse square-root singular stress components as the crack tip is approached. The amplitude parameters A_n are left undetermined in this local analysis and must be obtained from a complete solution of the problem; as in the case of the mode III problem, introducing the definition of the mode I *dynamic stress intensity factor*,

$$K_I = \lim_{\xi_1 \to 0} \sqrt{2\pi\xi_1} \, \sigma_{22}(r, 0^{\pm}) \tag{3.31}$$

the crack tip stress and displacement fields may be written as

$$\sigma_{\alpha\beta}(r, \theta) = \frac{K_I}{\sqrt{2\pi r}} f_{\alpha\beta}^I(\theta; v) + \sigma_{ox}(\alpha_d^2 - \alpha_s^2)\delta_{\alpha 1}\delta_{\beta 1} + \cdots$$

$$u_\alpha(r, \theta) = \frac{K_I\sqrt{r}}{\sqrt{2\pi}} g_\alpha^I(\theta; v) + \cdots \tag{3.32}$$

In the above, we have included the terms corresponding to $n = 1$ and 2 for stresses and $n = 1$ for displacements; the term corresponding to $n = 2$ implies a stress component parallel to the crack and is typically denoted by σ_{ox} in the literature on experimental investigations (Kobayashi and Mall, 1978) and is called the *T*-stress in the literature on quasi-static fracture (Cotterell and Rice, 1980). The functions $f_{\alpha\beta}^I(\theta; v)$ are given below

$$f_{11}^I(\theta; v) = \frac{1}{R(v)}\left\{(1 + \alpha_s^2)(1 + 2\alpha_d^2 - \alpha_s^2)\frac{\cos\frac{1}{2}\theta_d}{\gamma_d^{1/2}} - 4\alpha_d\alpha_s\frac{1}{\gamma_s^{1/2}}\cos\frac{1}{2}\theta_s\right\}$$

$$f_{22}^I(\theta; v) = \frac{1}{R(v)}\left\{-(1 + \alpha_s^2)^2\frac{\cos\frac{1}{2}\theta_d}{\gamma_d^{1/2}} + 4\alpha_d\alpha_s\frac{1}{\gamma_s^{1/2}}\cos\frac{1}{2}\theta_s\right\} \tag{3.33}$$

$$f_{12}^I(\theta; v) = \frac{2\alpha_d(1 + \alpha_s^2)}{R(v)}\left\{\frac{\sin\frac{1}{2}\theta_d}{\gamma_d^{1/2}} - \frac{\sin\frac{1}{2}\theta_s}{\gamma_s^{1/2}}\right\}$$

where

$$\gamma_d = \sqrt{1 - (v\sin\theta/C_d)^2} \quad \text{and} \quad \gamma_s = \sqrt{1 - (v\sin\theta/C_s)^2} \tag{3.34}$$

$$\tan\theta_d = \alpha_d\tan\theta \quad \text{and} \quad \tan\theta_s = \alpha_s\tan\theta \tag{3.35}$$

$$R(v) = 4\alpha_d\alpha_s - (1 + \alpha_s^2)^2 \tag{3.36}$$

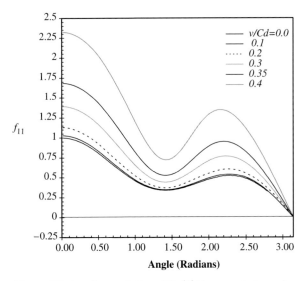

Figure 3.3 Angular variation of $f_{11}(\theta)$ for a mode I crack.

The angular variation of the functions $f_{11}^{\mathrm{I}}(\theta; v)$, $f_{22}^{\mathrm{I}}(\theta; v)$, and $f_{12}^{\mathrm{I}}(\theta; v)$, are shown in Figs. 3.3–3.5. The angular variation of the hoop component of the stress field is given in Fig. 3.6; this component was examined by Yoffe (1951). The shift in the peak from $\theta = 0°$ to $60°$ as the crack speed increased to about $v \sim 0.6C_{\mathrm{R}}$ was suggested as the cause of crack branching. The angular variation of the principal stress is shown in Fig. 3.7. It should be noted that the maximum principal stress component does not act

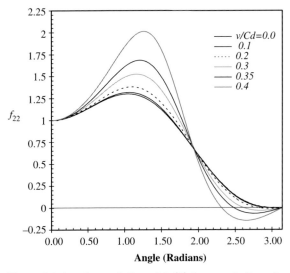

Figure 3.4 Angular variation of $f_{22}(\theta)$ for a mode I crack.

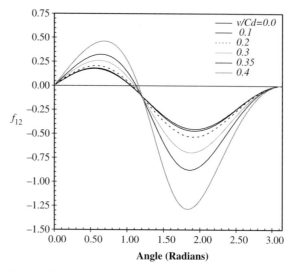

Figure 3.5 Angular variation of $f_{12}(\theta)$ for a mode I crack.

normal to the prospective crack line; Rice (1968) observed that the $f_{11}^I(\theta; v) > f_{22}^I(\theta; v)$ and hence paradoxical that the crack continues to grow along the x_1 direction. It can be seen from Eq. 3.29 that for $n = 2$ the only nonzero component is σ_{11} and is given by $3\mu A_2(\alpha_d^2 - \alpha_s^2)$, where A_2 is another constant to be determined from a complete analysis of the problem. However, if the crack surface is loaded by normal forces, the σ_{22} component corresponding to $n = 2$ will be equal to the crack face pressure.

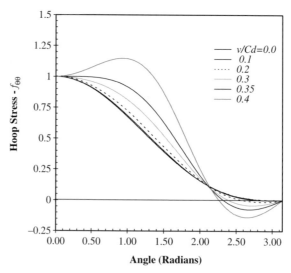

Figure 3.6 Angular variation of $f_{\theta\theta}(\theta)$ for a mode I crack.

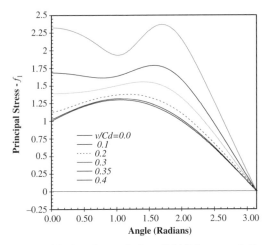

Figure 3.7 Angular variation of $f_1(\theta)$ for a mode I crack.

The displacement components corresponding to $n = 1$ are given by

$$
u_1(r, \theta) = \frac{2K_I}{\mu R(v)\sqrt{2\pi}} \left\{ (1 + \alpha_s^2) r_d^{1/2} \cos\left(\frac{\theta_d}{2}\right) - 2\alpha_d \alpha_s r_s^{1/2} \cos\left(\frac{\theta_s}{2}\right) \right\}
$$
$$
u_2(r, \theta) = \frac{2\alpha_d K_I}{\mu R(v)\sqrt{2\pi}} \left\{ (1 + \alpha_s^2) r_d^{1/2} \sin\left(\frac{\theta_d}{2}\right) - 2r_s^{1/2} \sin\left(\frac{\theta_s}{2}\right) \right\}
$$

(3.37)

It should be noted that the stress and displacement fields reduce to the appropriate quasi-static fields when $v \to 0$; however, the limit must be taken appropriately since $R(v)$ also tends to zero as $v \to 0$.

Freund (1990) has shown that if the crack moves with a nonuniform speed, the result described above carries over completely, with the only change that the stress intensity factor can now be considered to be a function of time and the instantaneous crack speed, $K_I(t, v)$. The effects of the nonuniform motion of the crack do not become apparent in the singular term or the constant term, but only in terms of higher order. In some experimental methods crack tip field information is extracted from distances that are far from the crack tip; in these applications, the higher order transient expansion is required to obtain an estimate of the stress intensity factor. We shall describe the transient field in Section 3.3.

3.2.3 In-Plane Antisymmetric Deformation

The stress and displacement field corresponding to a mode II crack growing at a speed v can be determined in a similar manner. If the antisymmetric solutions to Eq. 3.21 are used

$$
\varphi(r_d, \theta_d) = A r_d^\lambda \sin \lambda \theta_d
$$
$$
\psi(r_s, \theta_s) = B r_s^\lambda \cos \lambda \theta_s
$$

(3.38)

Then following the same procedure as described above, the mode II stress and displacement fields can be determined. They are given below

$$\sigma_{11}^n(r,\theta) = \frac{\mu A_n}{(1+\alpha_s^2)}\frac{n}{2}\left(1+\frac{n}{2}\right)\left\{(1+\alpha_s^2)(1+2\alpha_d^2-\alpha_s^2)r_d^{(n/2)-1}\right.$$

$$\left. \times \sin\left(\frac{n-2}{2}\theta_d\right) - \kappa(n)r_s^{(n/2)-1}\sin\left(\frac{n-2}{2}\theta_s\right)\right\}$$

$$\sigma_{22}^n(r,\theta) = \frac{\mu A_n}{(1+\alpha_s^2)}\frac{n}{2}\left(1+\frac{n}{2}\right)\left\{-(1+\alpha_s^2)^2 r_d^{(n/2)-1}\sin\left(\frac{n-2}{2}\theta_d\right)\right.$$

$$\left. +\kappa(n)r_s^{(n/2)-1}\sin\left(\frac{n-2}{2}\theta_s\right)\right\}$$

$$\sigma_{12}^n(r,\theta) = \frac{\mu A_n}{2\alpha_s}\frac{n}{2}\left(1+\frac{n}{2}\right)\left\{4\alpha_d\alpha_s r_d^{(n/2)-1}\cos\left(\frac{n-2}{2}\theta_d\right)\right.$$

$$\left. -\kappa(n)r_s^{(n/2)-1}\cos\left(\frac{n-2}{2}\theta_s\right)\right\}$$

(3.39)

where

$$\kappa(n) = \begin{cases} 4\alpha_d\alpha_s & \text{for } n \text{ even} \\ (1+\alpha_s^2)^2 & \text{for } n \text{ odd} \end{cases}$$

(3.40)

The corresponding displacement components are

$$u_1^n(r,\theta) = A_n\left(1+\frac{n}{2}\right)\left\{r_d^{n/2}\sin\left(\frac{n\theta_d}{2}\right) - \chi_1(n)r_s^{n/2}\sin\left(\frac{n\theta_s}{2}\right)\right\}$$

$$u_2^n(r,\theta) = \alpha_d A_n\left(1+\frac{n}{2}\right)\left\{r_d^{n/2}\sin\left(\frac{n\theta_d}{2}\right) - \chi_2(n)r_s^{n/2}\sin\left(\frac{n\theta_s}{2}\right)\right\}$$

(3.41)

The mode II dynamic stress intensity factor may be defined in analogy with the corresponding definition for the quasi-static problem; thus

$$K_{II} = \lim_{\xi_1\to 0}\sqrt{2\pi\xi_1}\,\sigma_{12}(r,0^{\pm})$$

(3.42)

For a general plane crack propagation problem, the crack tip stress field may then be represented as a superposition of the opening mode or mode I and the in-plane shearing mode or mode II dynamic stress field and characterized by the stress intensity factors K_I and K_{II}. The extent of the K-dominant field near the crack tip depends on the transient nature of the crack tip history; we shall explore this in Chapter 9.

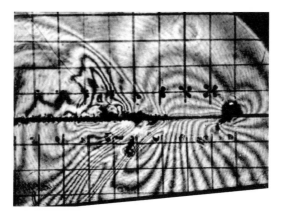

Figure 3.11 Isochromatic fringes at the tip of an intersonic crack $(C_s < v < C_d)$ in a homogeneous material. The crack was forced to grow as a shear crack by confining its path with a side-groove in the specimen. The shear Mach waves are clearly visible; curved shock indicates that the crack slowed down from a much greater speed. (Reproduced from Ravi-Chandar, 2001.)

see the recent article by Rosakis (2002). Kavaturu et al. (1998) have also examined intersonic crack growth. Ravi-Chandar (2001) demonstrated that intersonic crack growth is also possible in homogenous materials, without weak interfaces, if cracks are trapped along the maximum shear plane through other constraints such as a groove. Isochromatic fringe patterns corresponding to an intersonic crack within a groove in a Homalite-100 plate is shown in Fig. 3.11.

$U_-(\zeta, s) \sim 1/\zeta^{3/2}$ as $\zeta \to -\infty$. By Liouville's theorem, the bounded entire function must be a constant, which in this case is zero. Thus

$$\Sigma_+(\zeta) = \frac{\sigma^*}{\zeta}\left[\frac{F_+(0)}{F_+(\zeta)} - 1\right] = \frac{\sigma^*}{\zeta}\left[\frac{\sqrt{C_d(1-2\nu)/2}}{(1-\nu)}\frac{S_+(\zeta)(c+\zeta)}{\sqrt{a+\zeta}} - 1\right]$$

$$U_-(\zeta) = -\frac{b^2\sigma^*}{2\mu(b^2-a^2)}\frac{F_+(0)F_-(\zeta)}{\zeta}$$

(4.16)

The nontrivial task of inverting the transforms to determine the stress and displacements remains. In dynamic fracture problems of the kind considered here, the interest is primarily in the time dependence of the dynamic stress intensity factor; this can be extracted from the transformed solution in Eq. 4.16, without inverting the transform. From the Abel theorem, the dynamic stress intensity factor can be extracted from the asymptotic behavior of the Laplace transform $\hat{\sigma}_+(x_1, s)$ as $x_1 \to 0$

$$\lim_{x_1 \to 0} \sqrt{\pi x_1}\hat{\sigma}_+(x_1, s) = \lim_{\zeta \to \infty} \sqrt{s\zeta}\Sigma_+(s, \zeta)$$

(4.17)

Therefore, the Laplace-transformed dynamic stress intensity factor is given by

$$K_I(s) = \frac{\sqrt{2}\sigma^* F_+(0)}{s^{3/2}}$$

(4.18)

Now, the Laplace transform is easily inverted to yield the time-dependent dynamic stress intensity factor

$$K_I = \frac{2\sigma^*\sqrt{C_d t(1-2\nu)/\pi}}{(1-\nu)}$$

(4.19)

Eq. 4.19 provides the time variation of the dynamic stress intensity factor for the problem of stationary semi-infinite crack with uniform pressure loading on the crack surfaces, and the crack tip asymptotic field in Eq. 3.2 completes the characterization of the crack tip state.

The dynamic stress intensity factor result in Eq. 4.19 is applicable to the case when the crack surface pressure is imposed as a step function. In practical implementations of this kind of loading, the load is usually generated by a ramp-type loading with finite rise time. For this and other general time variations of the crack surface loading, the stress intensity factor may be determined using a superposition integral

$$K_I = \int_0^t K_I^{step}(t-\tau)\dot{f}(\tau)d\tau$$

(4.20)

where $\dot{f}(t)$ is the time derivative of the time variation of the applied load and $K_I^{step}(t)$ the stress intensity factor variation for the step loading. For example, let the applied crack surface pressure be a terminated ramp

$$f(t) = \begin{cases} \dfrac{t}{T} & 0 \leq t \leq T \\ 1 & t \geq T \end{cases}$$

(4.21)

where T is the rise time of the loading function. Then, from Eqs. 4.19 and 4.20, the dynamic stress intensity factor for the ramp loading problem is

$$
K_1(t) = \begin{cases} \dfrac{4}{3\pi} \dfrac{\sqrt{1-2\nu}}{1-\nu} \sigma^* \sqrt{\pi C_d T} \left(\dfrac{t}{T}\right)^{3/2} & 0 \leq t \leq T \\[4mm] \dfrac{4}{3\pi} \dfrac{\sqrt{1-2\nu}}{1-\nu} \sigma^* \sqrt{\pi C_d T} \left[\left(\dfrac{t}{T}\right)^{3/2} - \left(\dfrac{t}{T} - 1\right)^{3/2}\right] & t \geq T \end{cases} \tag{4.22}
$$

The dynamic stress intensity factor time histories for a step and terminated ramp loading are shown in Fig. 4.1. In some problems, it has been shown to be useful to obtain more information about the stress field than just the stress intensity factor. For this, the Laplace transforms must then be inverted. Inversion of the bilateral Laplace transform is accomplished by invoking the Cagniard-de Hoop technique. Formally, the inverse of the bilateral Laplace transform is written as

$$
\hat{\sigma}_+(x_1, s) = \frac{1}{2\pi i} \int_{\xi - i\infty}^{\xi + i\infty} \frac{1}{s^2} \Sigma_+(\zeta) e^{s\zeta x_1} s \, d\zeta \tag{4.23}
$$

where $-a < \xi < 0$. The path of integration is shown in Fig. 4.2. By completing the contour as indicated in the figure, the integral in Eq. 4.23 can be converted to an integral along the real axis. The integrand is analytic inside the closed contour and therefore by Cauchy's theorem, the integral is zero. However, by the decay of $\Sigma_+(\zeta)$ as $\zeta \to \infty$, it is evident that the integral along the circular arcs is zero. Therefore, Eq. 4.23 can be rewritten as an integral along the real axis

$$
\hat{\sigma}_+(x_1, s) = \int_{-a}^{-\infty} \frac{1}{\pi s} \text{Im}\{\Sigma_+(\zeta)\} e^{s\zeta x_1} \, d\zeta \tag{4.24}
$$

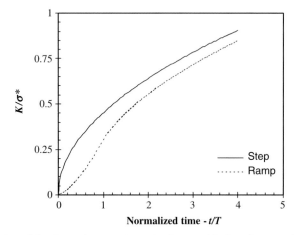

Figure 4.1 Variation of the dynamic stress intensity factor with time for a semi-infinite crack with pressure loading. Step and ramp loadings are shown.

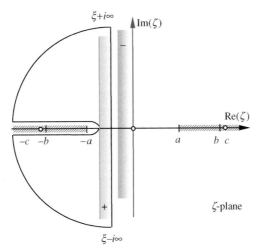

Figure 4.2 Complex ζ plane indicating the branch cuts and the contour of integration.

The Laplace transform with respect to time is inverted by rearranging the integral in such a manner that it represents the product of two Laplace transforms; the inversion is then immediate by the convolution theorem. This is accomplished by noting that Eq. 4.23 is a product of $1/s$ and an integral that can be recast in the form of a Laplace transform if ζx_1 is redefined as $-\eta$

$$\hat{\sigma}_+(x_1,s) = -\frac{1}{\pi s x_1} \int_{ax_1}^{\infty} \text{Im}\left\{ \Sigma_+\left(-\frac{\eta}{x_1}\right) \right\} e^{-s\eta} d\eta \qquad (4.25)$$

The inversion of the Laplace transform is then the following convolution integral

$$\sigma_+(x_1,t) = -\frac{1}{\pi x_1} \int_{ax_1}^{t} \text{Im}\left\{ \Sigma_+\left(-\frac{\eta}{x_1}\right) \right\} H(t - ax_1) d\eta \qquad (4.26)$$

Substituting for $\Sigma_+(\zeta)$ from Eq. 4.16 results in the following expression for the normal stress ahead of the crack line

$$\sigma_+(x_1,t) = -\frac{\sigma^*\sqrt{C_d(1-2\nu)/2}}{\pi(1-\nu)} \int_{ax_1}^{t} \frac{S_+(-\eta/x_1)(c-\eta/x_1)}{\eta\sqrt{a-\eta/x_1}} H(t - ax_1) d\eta \qquad (4.27)$$

Rescaling the variable of integration $\eta/ax_1 = \xi$, Eq. 4.27 can be rewritten in the following form

$$\sigma_+(x_1,t) = -\frac{\sigma^*\sqrt{C_d(1-2\nu)/2}}{\pi(1-\nu)} H(t/ax_1 - 1) \int_{1}^{t/ax_1} \frac{S_+(-\xi a)(c/a - \xi)}{\xi\sqrt{\xi-1}} d\xi \qquad (4.28)$$

Eq. 4.28 expresses the fact that there is neither a characteristic length nor a characteristic time in the problem and hence the fields must depend only on the ratio t/x_1. Eq. 4.28 was evaluated by Liu et al. (1998) in an effort to determine the build-up of the normal stress ahead of the crack tip for a stationary crack tip. Their calculation is displayed in Fig. 4.3; in this figure $\sigma_+(x_1,t)/\sigma^*$ is plotted as a function of $C_d t/x_1$. Consider a fixed position x_1 ahead of the crack tip; Fig. 4.3 shows the time evolution of the stress component σ_{22} at this point. Clearly, upon arrival of the dilatational wave at time $t = x_1/C_d$ (identified by the point A on the figure), a negative normal stress of increasing magnitude develops indicating that the initial wave to reach the point x_1 is a compressive wave. However, at $t = kx_1/C_d$ (identified by the point B on the figure) the shear wave from the crack tip arrives at the point x_1 and with it brings an increase in the normal stress; the normal stress does not become tensile until $t \sim 4x_1/C_d$. A finite rise time on the loading pulse as in Eq. 4.21 will increase this delay even more as discussed by Liu et al. (1998). Considering the terminated ramp loading as a sequence of steps, the normal stress ahead of the crack tip can be written as

$$\sigma_{22}(x_1,t) = \begin{cases} \dfrac{1}{T}\displaystyle\int_0^t \sigma_+\left(\dfrac{t-\tau}{ax_1}\right)d\tau & 0 < t < T \\[4mm] \dfrac{1}{T}\displaystyle\int_0^T \sigma_+\left(\dfrac{t-\tau}{ax_1}\right)d\tau & t > T \end{cases} \tag{4.29}$$

where $\sigma_+(\xi)$ is the stress field corresponding to the step loading determined in Eq. 4.28. This behavior has important consequences for crack initiation; if a stress-based fracture criterion is used—crack growth will begin when the normal stress σ_{22} at a certain distance, r_c, from the crack tip reaches a critical fracture stress, σ_f—then the crack cannot initiate at least until after $t \geq 4r_c/C_d$. It has been suggested by Liu et al. (1998) that this leads to a loading rate dependence on crack initiation even in linearly elastic, brittle materials as described in Chapter 10.

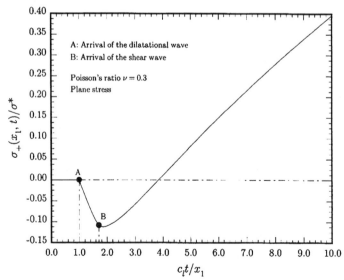

Figure 4.3 Variation of $\sigma_+(x_1,t)$ ahead of the crack tip for a semi-infinite crack under uniform pressure loading. (Reproduced from Liu et al., 1998.)

4.1.2 Semi-Infinite Crack Under a Point Load

We now consider a problem involving a characteristic length; consider again an unbounded linearly elastic body containing a semi-infinite crack lying along the negative x_1 axis with its tip at $x_1 = 0$. The loading is prescribed as a point force of unit magnitude applied at some distance l behind the crack tip as illustrated in Fig. 4.4; the corresponding boundary conditions are as follows

$$
\begin{aligned}
\sigma_{22}(x_1, 0, t) &= \sigma_+(x_1, t) - \delta(x_1 + l)H(t) \quad -\infty < x_1 < \infty \\
\sigma_{12}(x_1, 0, t) &= 0 \quad -\infty < x_1 < \infty \\
u_2(x_1, 0, t) &= u_-(x_1, t) \quad 0 < x_1 < \infty
\end{aligned}
\tag{4.30}
$$

where the domain for $\sigma_+(x_1, t)$ is $x_1 > 0$ and the domain for $u_-(x_1, t)$ is $x_1 < 0$, but both are unknown functions to be determined. If the procedure outlined in the previous section is followed, an equation analogous to the Wiener-Hopf equation in Eq. 4.7 is obtained

$$
-\mu s \frac{\mu R(\zeta)}{b^2 \alpha(\zeta)} U_-(\zeta, s) = \Sigma_+(\zeta, s) - \frac{1}{s} e^{ls\zeta}
\tag{4.31}
$$

The Rayleigh function is factorized just as indicated in Eq. 4.11; then, the above equation can be rewritten as

$$
-\frac{2\mu s(b^2 - a^2)}{b^2} \frac{U_-(\zeta, s)}{F_-(\zeta)} = \Sigma_+(\zeta, s)F_+(\zeta) - \frac{1}{s} e^{ls\zeta} F_+(\zeta)
\tag{4.32}
$$

Unlike the problem in the previous section, the transform of the loading on the right hand side is not of a form where the dependence on s and ζ may be separated. To circumvent this difficulty, Freund (1974b) used a superposition scheme. First, he determined the stress intensity factor corresponding to a moving dislocation corresponding to the following boundary conditions

$$
\begin{aligned}
\sigma_{22}(x_1, 0, t) &= \sigma_+(x_1, t) \quad -\infty < x_1 < \infty \\
\sigma_{12}(x_1, 0, t) &= 0 \quad -\infty < x_1 < \infty \\
u_2(x_1, 0, t) &= u_-(x_1, t) + H(vt - x_1) \quad 0 < x_1 < \infty
\end{aligned}
\tag{4.33}
$$

Then, superposing solutions to this problem, the displacements generated in the point load problem (the Lamb problem discussed in Chapter 2) on $x_1 > 0$ were negated. The dynamic

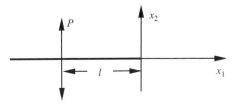

Figure 4.4 Semi-infinite crack with a point force of magnitude P applied at a distance l behind the crack tip.

stress intensity factor was then extracted. Kuo and Chen (1992) re-examined this problem recently and provided a solution by factoring the last term on the right hand side directly. After defining the branches of the function $F_\pm(\zeta)$ exactly as in Section 4.1.1, it is apparent that the left hand side of Eq. 4.32 is analytic in $\text{Re}(\zeta) < 0$ and the first term on the right hand side of Eq. 4.32 is analytic in $\text{Re}(\zeta) > -a$. However, the exponential term makes the right hand side become unbounded as $\zeta \to \infty$ and hence this term must be split into two bounded analytical functions; this was accomplished by Kuo (1993). From this point onwards, the procedure for extraction of the stress intensity factor is identical to that described in the previous section. The analytic continuation arguments described in connection with the previous problem can again be applied to show that

$$\Sigma_+(\zeta, s) = \frac{1}{s} \frac{S_+^*(\zeta, s)}{F_+(\zeta)} \tag{4.34}$$

where

$$S_+^*(\zeta, s) = F_+(\zeta)e^{ls\zeta} - S_-^*(\zeta, s) = -\frac{1}{2\pi i} \int_{Br} \frac{e^{ls\xi}\sqrt{a+\xi}}{(\xi+c)S_+(\xi)} \frac{d\xi}{\xi-\zeta} \tag{4.35}$$

and Br stands for the Bromwich contour. The transform of the dynamic stress intensity factor is obtained by applying the Abel theorem, as in Eq. 4.17; inverting this, the dynamic stress intensity factor for the point load problem posed in Eq. 4.29 is then found to be

$$K_I^F(t) = \begin{cases} -\dfrac{2}{\pi}\sqrt{\dfrac{2}{\pi l}}\left(\dfrac{b^2}{a^2} - 1\right)\displaystyle\int_1^{t/l} \phi(\xi)\dfrac{d\xi}{\sqrt{t/l-\xi}} & a < \dfrac{t}{l} < b \\[3mm] \sqrt{\dfrac{2}{\pi l}}\left[1 - \dfrac{\sqrt{c-a}}{(c-t/l)S_-(c)}\right] & b < \dfrac{t}{l} < c \\[3mm] \sqrt{\dfrac{2}{\pi l}} & \dfrac{t}{l} > c \end{cases} \tag{4.36}$$

where

$$\phi(\xi) = \frac{\sqrt{\xi-1}(c/a - \xi)(2\xi^2 - b^2/a^2)S_+(a\xi)}{[16\xi^2(\xi^2-1)(b^2/a^2 - \xi^2) + (2\xi^2 - b^2/a^2)^4]}$$

$K_I^F(t)$ is the dynamic stress intensity factor per unit load; the superscript F is used to indicate that this solution may be used as a fundamental solution for application in superposition schemes. The stress intensity factor variation with time is shown in Fig. 4.5. Upon arrival of the dilatational wave, a negative stress intensity of small magnitude develops indicating the compressive stress from this wave. Upon arrival of the shear wave, this compression increases without bound, and after passage of the Rayleigh wave, the stress intensity factor settles down at the value equal to the static value. Clearly, the negative stress intensities are not physical and should not appear in sharp cracks. However, in machine-notched cracks, this would imply a closing of the gap.

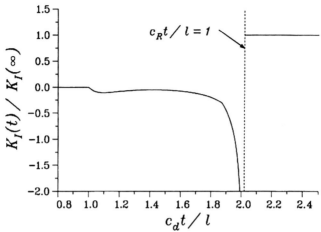

Figure 4.5 **Time variation of the dynamic stress intensity factor for a point load on a half space at $x_1 = -l$. (Reproduced from Freund, 1990.)**

With the above solution, it is now possible to generate the dynamic stress intensity factors for semi-infinite cracks with arbitrary load variations along the crack surface through superposition. Consider a semi-infinite crack with a non-uniform load distribution behind the crack tip as shown in Fig. 4.6. The dynamic stress intensity factor may then be calculated as

$$K_I(t) = \int p(\xi) K_I^F(t, \xi) d\xi \tag{4.37}$$

where the range of integration is over the load distribution.

Consider as an example, the loading condition where a uniform pressure loading is applied over a length L behind the crack tip

$$p(x_1, t) = -p_0 H(x_1 + L) H(t) \quad -\infty < x_1 < 0 \tag{4.38}$$

Introducing this loading in Eq. 4.37, the stress intensity factor is obtained as

$$K_I(t) = 2p_0 \sqrt{2L/\pi} \quad \text{for } t > cL \tag{4.39}$$

Note that while the dynamic stress intensity factor for $aL < t < cL$ may be obtained, only the range given in Eq. 4.39 is of interest in most problems. Solutions to other problems may be constructed by superposition; e.g., consider a semi-infinite crack that is *unloaded*

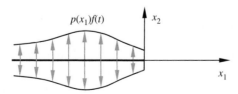

Figure 4.6 Semi-infinite crack with a distributed force behind the crack tip.

over a length L near the crack tip; the stress intensity factor is obtained by the superposition of Eqs. 4.19 and 4.39

$$K_1(t) = 2\sigma^* \left[\frac{\sqrt{C_d t(1 - 2\nu)/\pi}}{(1 - \nu)} - \sqrt{\frac{2L}{\pi}} \right] \quad \text{for } t > cL \tag{4.40}$$

Comparison of these theoretical estimates with experimental measurements is discussed in Chapter 9.

4.2 Analysis of Moving Crack Problems

The asymptotic field near moving cracks discussed in Section 3.2 provides the basis for characterizing stress field in terms of a single parameter—*the dynamic stress intensity factor*. We now turn to an evaluation of the stress intensity factor for moving crack problems. We shall first discuss an idealized problem considered by Yoffe, and then discuss a general method for the determination of the stress intensity factor for arbitrary loading.

4.2.1 The Yoffe Problem

The first analytical solution to the problem of a moving crack was provided by Yoffe (1951), who considered the problem of a crack of constant length moving along a straight line in an infinite two-dimensional medium under remote tractions, symmetric with respect to the crack. This is not a physically realistic situation since it requires the crack to open and grow at one end, but close or heal at the other. However, since the elastodynamic equations are usually solved after decoupling them from the failure criterion, the structure of the crack tip stress field is still appropriate as we shall observe later. Yoffe examined the asymptotic field to identify possible criteria for crack branching; we shall discuss this aspect in Chapter 11. The solution to the Yoffe problem is presented below, but assuming a mode II loading instead of Yoffe's original mode I problem. The solution procedure follows the development of Freund (1990) for the mode I problem. Broberg (1999) has presented a solution of the mode II Yoffe problem by requiring the stress intensity factor at the trailing or healing crack tip to go to zero; this model is believed to be a physically realistic picture of slip events in sliding interfaces such as in earthquakes.

Consider an infinite body with stresses decaying to zero as $r = \sqrt{x_1^2 + x_2^2} \rightarrow \infty$. A crack of length $2a$ is assumed to lie initially along $x_2 = 0$, and to move at a *constant* speed $v < C_R$. It is under a uniform stress τ^∞ along the crack faces as indicated in Fig. 4.7; it is possible to convert this problem to one where the crack surfaces are traction-free and the remote loading is a uniform shear, simply by adding a uniform shear τ^∞ over the entire body. Introducing the Galilean transformation $\xi_1 = x_1 - vt$, $\xi_2 = x_2$, the governing differential equations for the potentials $\varphi(x_1, x_2, t)$ and $\psi(x_1, x_2, t)$ can be rewritten in terms of the distorted polar coordinates as

$$\nabla^2 \varphi(r_d, \theta_d) = 0, \quad \nabla^2 \psi(r_s, \theta_s) = 0 \tag{4.41}$$

result already displayed in Eq. 4.56. The dynamic stress intensity factor for a crack propagating at an arbitrary speed v can be written as the product of a universal function of the crack speed and the instantaneous stress intensity factor of a stationary crack of equivalent crack length

$$K_1(t, l(t), v) = k(v)K_1^0(t, l(t), 0) \tag{4.74}$$

where $K_1^0(t, l, 0)$ is the stress intensity factor corresponding to a stationary crack of length l at time t.

The result above was obtained for a semi-infinite crack in an unbounded medium; however, as Freund (1990) suggests this result can be generalized to any complex geometrical and boundary conditions through a superposition of solutions to problems involving semi-infinite crack problems. For finite crack problems, the result remains valid until the waves from one crack tip arrives at the other crack tip as demonstrated by Kostrov (1975). A similar result may be derived for the problem of mode II crack growth; the expression corresponding to Eq. 4.74 is obtained by replacing the dilatational wave speed by the distortional wave speed.

As an example of the application of this method of determining the dynamic stress intensity factor, consider a semi-infinite crack, loaded by a step tensile stress wave of amplitude σ^* as illustrated in Fig. 4.9. When the stress wave reaches the crack plane at time $t = 0$, the traction-free boundary condition on this plane would require that a uniform compression be generated on the crack plane, regardless of whether the crack is stationary or moving. For a stationary crack, this corresponds to the uniformly pressure-loaded semi-infinite crack considered in Section 4.1 and the stress intensity factor is therefore given in Eq. 4.19. If the crack begins to grow at a speed v at some time $t = \tau$, the stress intensity factor is still given by Eq. 4.19, but with the multiplying factor $k(v)$; the complete stress

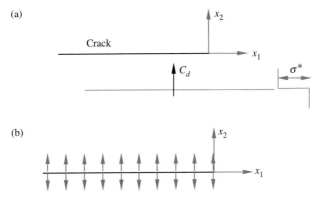

Figure 4.9 (a) Stress pulse incident on a semi-infinite crack. (b) Equivalent semi-infinite crack with uniform pressure loading on the crack surfaces.

intensity factor history for a growing crack can be written as

$$K_1 = \begin{cases} \dfrac{2\sigma^* \sqrt{C_d t(1-2v)/\pi}}{(1-v)} & \text{for } 0 < t < \tau \\[4mm] k(v)\dfrac{2\sigma^* \sqrt{C_d t(1-2v)/\pi}}{(1-v)} & \text{for } t > \tau \end{cases} \tag{4.75}$$

Note that there is an abrupt drop in the stress intensity factor at crack initiation associated with the sudden transition from a stationary crack to a crack moving at a speed v.

Next, consider that the loading is actually applied on the crack plane rather than by an impinging stress wave; then, as the crack extends, the newly created crack must be traction free; thus

$$\begin{aligned} \sigma_{22}(x_1, 0^+, t) &= -\sigma^* \quad \text{for } -\infty < x_1 < 0 \\ \sigma_{22}(x_1, 0^+, t) &= 0 \quad \text{for } 0 < x_1 < l(t) \\ u_2(x_1, 0^+, t) &= 0 \quad \text{for } x_1 > l(t) \\ \sigma_{12}(x_1, 0^+, t) &= 0 \quad \text{for } -\infty < x_1 < \infty \end{aligned} \tag{4.76}$$

Let the crack be stationary for time $t < \tau$ and further, let the crack extend at a constant speed, v along the x_1 axis. This problem is analogous to that considered at the end of Section 4.1.2. We now have to superpose the solution to a semi-infinite pressure-loaded crack to that where the pressure loading is applied only over the newly created crack surface, thus creating a traction free crack plane. Thus, we superpose the solution in

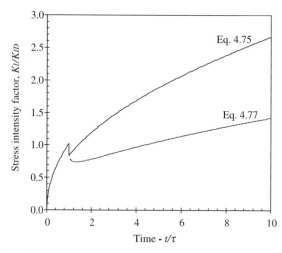

Figure 4.10 Variation of the dynamic stress intensity factor (normalized by its value at $t = \tau$) with time (normalized by the time at initiation, τ) for a crack growing at a constant speed ($v/C_d = 0.1$, $v = 0.3$).

Eq. 4.19 to the solution in Eq. 4.40, letting $L = v(t - \tau)$. In addition, the multiplicative factor $k(v)$ must be introduced for the growing crack

$$
K_1 = \begin{cases}
2\sigma^* \dfrac{\sqrt{C_d t(1 - 2v)/\pi}}{(1 - v)} & t < \tau \\[4mm]
2\sigma^* k(v) \left[\dfrac{\sqrt{C_d t(1 - 2v)/\pi}}{(1 - v)} - \sqrt{\dfrac{2v(t - \tau)}{\pi}} \right] & t > \tau
\end{cases}
\tag{4.77}
$$

Note again that there is an abrupt drop in the stress intensity factor at crack initiation. A sketch of the time variation of the dynamic stress intensity factor for stress pulse loading and for the case of crack face loading over the original crack surfaces is shown in Fig. 4.10. Thus, the main advantage of Eq. 4.74 is obvious; we can take any known solution for a stationary crack and then determine the stress intensity factor variation for a dynamically growing crack. In the next chapter we examine the results of an experimental implementation of the problem.

It is important to recognize that what we have at this stage are dynamically admissible solutions; given a uniform pressure loading on the original semi-infinite crack surface and given that crack initiation occurs at $t = \tau$, and furthermore, that the crack grows at a constant speed v, Eq. 4.74 yields the time variation of the dynamic stress intensity factor. A criterion for the dissipation must be imposed to establish the conditions at crack initiation and for establishing the crack tip equation of motion. However, from the nature of the function $k(v)$, it is clear that the dynamic stress intensity factor for a growing opening mode crack will vanish as the crack speed approaches the Rayleigh wave speed and thus from the elastodynamic point of view, such cracks cannot propagate at a speed larger than the Rayleigh wave speed.

propagation load or at the final displacement. None of these estimates is equal to the Charpy V-notch energy, C_v, which measures the difference between the potential energy difference in the striker before and after impact. It is not clear how the energies calculated above may be used; it appears that the instrumented drop-weight simply provides a different measure of impact toughness that the Charpy tests, but not readily interpretable in terms of the Charpy V-notch energy.

On the other hand, the measurements from the drop-weight or instrumented impact test may be used to determine the dynamic crack initiation toughness. However, in order to accomplish this, one must determine the dynamic stress intensity factor at initiation of crack growth. Different methods have been used for the calculation of the dynamic stress intensity factor. Could the dynamic stress intensity factor be determined from the quasi-static analysis of the three-point bend configuration with the dynamically measured load? Ireland (1974) indicated that this procedure could be used when the time to fracture t_f was longer than about three times the time of the characteristic oscillation, τ. The rationale is that after about the third oscillation, the amplitude of the load oscillations becomes small enough to be neglected. Thus, the dynamic stress intensity factor is estimated as

$$K_{\mathrm{I}}^{\mathrm{stat}}(t) = \frac{P(t)S}{BW^{3/2}} f\left(\frac{a}{W}\right), \quad t > 3\tau \tag{6.3}$$

where a is the crack length, S the span of the beam, W the specimen width, and

$$f(\alpha) = \frac{3\sqrt{\alpha}[1.99 - \alpha(1 - \alpha)(2.15 - 3.93\alpha + 2.7\alpha^2)]}{2(1 + 2\alpha)(1 - \alpha)^{3/2}}$$

is the geometric factor for the three-point bend specimen given by Srawley (1976). Therefore, in the experiment, if the time for crack initiation and the tup load are measured independently, the dynamic crack initiation toughness can be calculated through Eq. 6.3. However, it is possible to induce crack initiation for times $t_f < 3\tau$ simply by altering the specimen dimensions or the impact speed. Under this condition, one must resort to a dynamic analysis. Two approaches have been used to determine the dynamic stress intensity factor: impact response curves (Kalthoff, 1990a) and key curves (Böhme, 1990).

Impact response curves are in fact calibration curves; the basic idea is that the true dynamic time variation of the stress intensity factor for a stationary crack is determined through analysis, numerical simulation or experimental measurements (using one of the methods described in Chapter 8)—this is the *impact response curve*. The crack is forced to remain stationary in this test by making a blunt initial crack. Then, to measure the dynamic crack initiation toughness, a fatigue-precracked specimen is then impacted at the same speed in the same geometry. Measuring only the time to initiate the crack, the dynamic stress intensity can be read-off from the impact response curve.

The concept of a key curve was introduced by Böhme (1990); in this approach, first, the impact event is modeled with a spring-mass model—the tup of mass m and the specimen of compliance C_{sp} (the support compliance C_{sup} may also be included) constitute the spring mass system. An estimate of the stress intensity factor, $K_{\mathrm{I}}^{qs}(t)$, is obtained from this model. The dynamic stress intensity factor is then written in terms of this estimate

$$K_I^{\text{dyn}}(t) = k_{\text{key}}(t)K_I^{qs}(t) \tag{6.4}$$

where $k_{\text{key}}(t)$ is called the *dynamic key curve*; it must be determined by a comparison of $K_I^{qs}(t)$ to $K_I^{\text{dyn}}(t)$ obtained from analysis, numerical simulation or experimental measurements. Böhme (1990) made a comparison of the three methods described above—static analysis, key curves and direct measurements of the dynamic stress intensity factor; his result is reproduced in Fig. 6.3. The optical method of caustics (described in Section 8.3) was used to determine the dynamic stress intensity factor. $K_I^{\text{dyn}}(t)$ exhibits a small amplitude oscillation about a mean curve; the mean curve is well estimated by the quasi-static analysis suggesting that the spring-mass model is a reasonably good approximation to the actual dynamic variation. However, if one is interested in characterizing critical material properties, the discrepancy between $K_I^{qs}(t)$ and $K_I^{\text{dyn}}(t)$ is not acceptable. Also, the static estimate, $K_I^{\text{stat}}(t)$, based on Eq. 6.3 is quite different from the actual variation whenever the time to failure is $t_f < 3\tau$. Kalthoff (1990a) shows examples of impact tests where the static analysis is inadequate even for $t_f > 3\tau$.

From the wealth of test results that have been obtained on drop-weight or instrumented impact tests, it is clear that a dynamic interpretation—analytical, numerical or experimental—is necessary to evaluate stress intensity factor from impact tests. Thus, both the key curve and impact response curve concepts are useful in evaluating data from

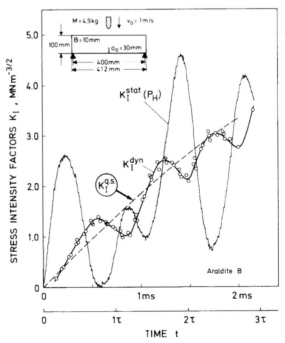

Figure 6.3 Comparison of stress intensity factors for an impact test estimated using static, quasi-static and dynamic analysis. (Reproduced from Böhme, 1990.)

impact tests. Two additional concerns arise in using these curves: first, the nature of dynamic contact between the specimen and the tup is difficult to control between different experiments. Second, only the load at the tup is measured, with the implicit assumption that the contact between the specimen and anvil is reproducible. In fact, there is loss of contact between the specimen and anvil (Kalthoff, 1990a) which makes the analysis less reliable. Hence, the procedures for the interpretation of drop-weight and instrumented impact tests are far from satisfactorily established. Direct measurement of the dynamic stress intensity factor in each test through one of the techniques discussed in Chapter 8 is the best strategy for the unambiguous evaluation of the dynamic stress intensity factor from the impact tests.

6.3 Projectile Impact

Projectile impact experiments are just variations of the drop-weight or other instrumented impact tests; once again, specimens of different geometrical configurations are impacted by a projectile of mass m moving at a speed v. The only difference in these impact experiments is that the projectile (weighing anywhere from a few grams to a few kilograms) is launched from the barrel of a gun at speeds in the range of a 10 m/s to 1 km/s; the higher speeds lead to penetration, a much more complex problem of dynamic failure. For characterization of dynamic fracture, usually projectile speeds in the range of 10–200 m/s is used (Kalthoff, 1990a; Ravichandran and Clifton, 1989; Taudou et al., 1992; Mason et al., 1992 and many others). The loading duration is governed by the length of the projectile and is typically much shorter than in the drop-weight test—about 10–200 μs. The time to fracture is typically in the range of 1–100 μs. With such short duration tests, dynamic analysis of the response of the specimen is clearly necessary for the interpretation of the test results; in many cases, crack initiation begins even before stress waves from the impact reach all the far boundaries of the specimen. Loading generated by such projectile impact onto a specimen can yield loading rates in the range of $\dot{K}_I^{dyn}(t, v) = 10^5$ MPa$\sqrt{m}$/s at the low impact speeds to about $\dot{K}_I^{dyn}(t, v) = 10^8$ MPa$\sqrt{m}$/s at the high projectile speeds for a very short duration (less than 1 μs). Apparatus for generating such impact loads are not commercially available and must be designed and fabricated especially for each implementation. But with this wide range of loading rates, projectile impact tests are suitable for characterization of the loading rate dependence of the dynamic crack initiation toughness. Because the projectile velocity can be controlled to within close tolerances, this method of loading is quite repeatable.

The projectiles used are typically circular cylinders with a flat nose; impact of the flat nose on the edge of the specimen generates a compression wave in the specimen that travels through the specimen at the speed C_d. The magnitude of the stress wave depends on the impedance mismatch between the projectile and the specimen. The duration of the loading pulse depends on the length of the projectile. If the impedance of the projectile and the specimen are matched, upon impact a compression wave of equal magnitude travels both into the specimen and into the projectile. The magnitude of the compressive stress may be estimated from one-dimensional wave theory to be $\sigma^* = v\sqrt{\rho E}/2$ where ρ is the density, and E the modulus of elasticity, but since the specimen is usually larger than the diameter of the projectile, this estimate of magnitude is unlikely to be accurate far from the impact point. The tensile pulse reflected from the far end of the projectile cannot pass through the impact surface and the loading of the specimen ends. Thus, the duration of the loading pulse is $\tau = 2L_{\mathrm{p}}/C_0^{\mathrm{proj}}$, where L_{p} is the length of the projectile and C_0^{proj} is the bar wave speed in the projectile. The compressive pulse traveling in the specimen can be used in different ways to generate a transient load history on the cracked specimen. Three different implementations are shown in the sketches in Fig. 6.4. In the first implementation shown in Fig. 6.4a, the projectile is made to impact opposite to the side containing the edge crack. While this might appear to be similar to the 'three-point bend' specimen we refrain from using this terminology; the specimen is usually not supported by rigid supports or anvils, but is simply suspended by strings. The compressive stress wave travels to the free boundary and reflects as a tensile wave. When this wave reaches the crack tip, the mode I stress intensity factor grows until crack initiation and growth are generated. The second configuration is shown in Fig. 6.4b where the projectile travels in the direction normal to the crack line; once again, impact at the top end of the specimen generates a compression wave that travels to the stress free boundary at the bottom, reflects as a tensile pulse and loads the crack in the middle of the specimen with a mode I loading symmetry. Closed form solutions of the elastodynamic problem for calculation of the dynamic stress intensity factor are not available for these configurations even for the case of stationary cracks except for very short times. Consider a tensile stress pulse of magnitude σ^* and duration τ propagating with a speed C_{d} as indicated schematically in Fig. 6.5. Let the stress wave reach the crack tip at time $t = 0$. Before waves from the corner A in Fig. 6.5 arrive at the crack tip—i.e. for $t < a/C_{\mathrm{d}}^p$ the mode I stress intensity factor is given by (Freund, 1990):

$$K_{\mathrm{I}}^{\mathrm{dyn}}(t) = 2\sigma^* \frac{\sqrt{C_{\mathrm{d}}(1 - 2\nu)/\pi}}{1 - \nu} \sqrt{t} \quad \text{for } t < \tau$$

$$K_{\mathrm{I}}^{\mathrm{dyn}}(t) = 2\sigma^* \frac{\sqrt{C_{\mathrm{d}}(1 - 2\nu)/\pi}}{1 - \nu} (\sqrt{\tau} - \sqrt{t - \tau}) \quad \text{for } \tau < t < a/C_{\mathrm{d}}$$

(6.5)

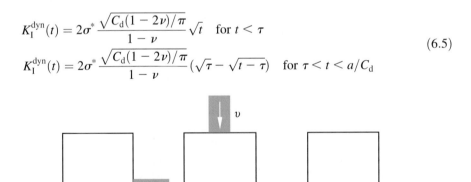

(a)   (b)   (c)

Figure 6.4 Configurations for projectile impact. (a) Impact parallel to the crack line. (b) Impact normal to the crack line. (c) Impact parallel to the crack line, but just below the crack line. The specimens are usually not supported rigidly, but simply suspended with strings. The dimensions of the specimen can be varied over quite a large range in experimental implementations.

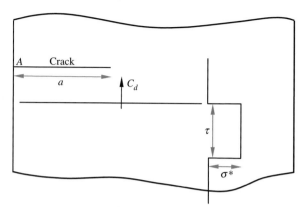

Figure 6.5 Interaction of a tensile stress pulse magnitude σ^* and duration τ with a crack. The stress pulse travels with a speed C_d.

It is assumed that the duration of the loading pulse is shorter than the time for arrival of the wave from the corner: $\tau < a/C_\mathrm{d}$. If this is not the case, the stress intensity factor for times $0 < t < a/C_\mathrm{d}$ is given by the first expression in Eq. 6.5. For $t > a/C_\mathrm{d}$, the waves from the corner A must be taken into account and this makes analysis very difficult; then, one must resort to numerical simulations using finite element analysis or to direct experimental evaluation through the methods discussed in Chapter 8. In practice, departures from the ideal impact conditions described here such as finite rise times of the loading pulse, variations in the impedance between the projectile and the specimen, and departure from the one-dimensional theory considered here will force the need for numerical simulations or direct experimental measurements even for short times. While both the configurations in Fig. 6.4a and b generate a mode I loading on the crack, the main difference between the two is in how the nonsingular stress field in the neighborhood of the crack develops. This is known to influence the crack growth direction (Cotterell and Rice, 1980), but its influence on crack initiation is not well established.

Finally, in the configuration shown in Fig. 6.4c, the projectile is made to impact the specimen just below the line of the crack. The compression wave generated from this impact travels along the bottom half of the specimen and diffracts around the crack; points below the crack line are made to displace to the right while points above are at rest suggesting that the loading is anti-symmetric. This loading method generates a mode II dynamic loading at the crack tip and has been used for evaluating the dynamic response of cracks under anti-symmetric loading (Kalthoff, 1990b; Mason et al., 1992; Ravi-Chandar, 1995). Lee and Freund (1990) determined the stress intensity factor variation with time for this loading configuration

$$K_\mathrm{I}^\mathrm{dyn}(t) = \frac{Ev}{2C_\mathrm{d}} \sqrt{\frac{a}{\pi}} g(t), \ K_\mathrm{II}^\mathrm{dyn}(t) = \frac{Ev}{2C_\mathrm{d}} \sqrt{\frac{a}{\pi}} h(t) \quad \text{for } t < \sqrt{a^2 + d^2}/C_\mathrm{d} \quad (6.6)$$

where d is the diameter of the projectile, E the modulus of elasticity of the specimen, and $g(t)$ and $h(t)$ are functions defined by Lee and Freund (1990). For this loading, the mode I stress intensity factor turns out to be negative, which implies contact between the top and bottom crack surfaces; Eq. 6.6 may be used only when the initial crack is a blunt crack with a large opening that prevents such contact. For $t > \sqrt{a^2 + d^2}/C_d$, the finite diameter of the projectile brings about differences in the loading from that considered in the analysis and the analytical results are no longer applicable. Once again recourse must be made to numerical simulation or experimental measurements as in the work of Mason et al. (1992). In mode II experiments such as this, it is possible to determine the critical condition for crack initiation; in some cases, an opening mode crack is initiated at an angle to the initial crack while in the other a shear crack or a shear band extends along the extension of the original crack line; these problems are not addressed in this book.

One final variation of the projectile impact loading scheme merits additional description—the plate impact test. In this test, both the projectile and the specimen are extremely thin disks as illustrated in Fig. 6.6. In this figure, the projectile, called the 'flyer plate' is propelled from a gas gun, usually mounted on a hollow fiber glass carrying tube, at speeds of up to about 1 km/s. This flyer plate impacts the thin-disk specimen that contains a crack across half of its cross-section; the arrangement is very similar to that illustrated in Fig. 6.4b, except that the cross-section of the projectile and specimen are identical here. The pulse duration $\tau = 2L_p/C_0^{proj}$ is extremely short because the projectile length is small—for example, $L_p = 5$ mm, $C_0^{proj} = 5000$ m/s results in a pulse of duration $\tau = 1$ μs. Details of how these are implemented for the mode I loading are given in Ravichandran and Clifton (1989). The loading rate achieved in these experiments is about $\dot{K}_I^{dyn}(t, v) = 10^8$ MPa√m/s. While the loading scheme generates a well-characterized one-dimensional pulse loading, the diagnostics of the specimen response is quite difficult; the normal, and tangential displacements of the specimen backplane are monitored with

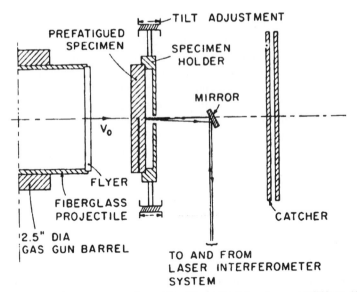

Figure 6.6 Plate impact test configuration. (From Ravichandran and Clifton, 1989.)

the aid of interferometric velocimeters and interpreted in terms of the dynamic stress intensity factor. The alignment and set up of this apparatus are quite challenging tasks.

6.4 Hopkinson Bar Impact Test

The Hopkinson pressure bar is a one-dimensional wave guide, usually of circular cross-section. One-dimensional compression stress wave of magnitude $\sigma^* = v\sqrt{\rho E}/2$ and duration $\tau = 2L_p/C_0^{proj}$ can be set up in this bar by projectile impact as discussed in Section 6.3. It is also possible to generate tensile waves by suitable arrangement of impact or through the use of explosives. Many variations of the Hopkinson bar loading scheme appear in the literature; for a detailed description of Hopkinson bar procedures, see the ASM Metals Handbook, Volume 8.

Three versions of this method are illustrated in Fig. 6.7. The simplest arrangement is that used by Costin et al. (1977); as illustrated in Fig. 6.7a, the specimen is a long round bar, B, with a circumferential precrack introduced by fatigue cycling. An explosive charge is detonated at one end of the specimen resulting in a tensile pulse traveling down the length of the specimen. Strain gages are placed on either side of the crack to monitor the stress waves on either side of the crack; in addition, Costin et al. (1977) used a moiré technique to determine the crack opening displacement. The rise time of the loading pulse was in the range of 35–40 μs, and fracture occurred within 20–25 μs. The type of experimental measurement obtained in this apparatus is shown in Fig. 6.8, where the time variation of the measured load and crack opening displacement are shown. It should be noted that the quality of signals is extremely good in comparison to the noisy signals obtained in an instrumented impact test. From such measurements, the dynamic stress intensity factor may be estimated. Costin et al. (1977) used this method to determine

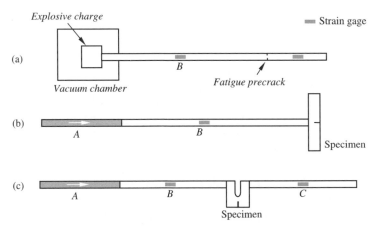

Figure 6.7 Hopkinson pressure bar arrangements for fracture tests. (a) Tension pulse generated by an explosive charge (Costin et al., 1977). (b) Impact generated on single edge notched specimen (Nicholas, 1975; Ruiz and Mines, 1985). (c) Compact compression specimen (Maigre and Rittel, 1995).

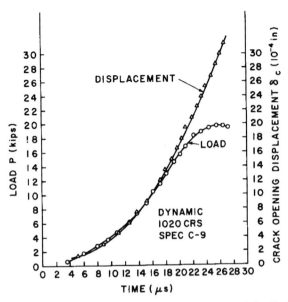

Figure 6.8 Time variation of the load measured with the strain gage and the displacement measured with the moiré technique. (Reproduced from Costin et al., 1977.)

the rate dependence of the dynamic crack initiation toughness at loading rates of the order of $\dot{K}_{\mathrm{I}}^{\mathrm{dyn}}(t, v) = 10^6$ MPa$\sqrt{\mathrm{m}}$/s.

The second implementation of the Hopkinson bar shown in Fig. 6.7b is a variation of the instrumented impact test; instead of relying on a falling weight or a projectile, a well-defined one-dimensional compression stress wave is generated by the projectile A impacting the bar B; this wave travels down the length of the bar and impacts on a single-edge-notched specimen as in the typical instrumented impact test (Nicholas, 1975). The strain gage signal can be interpreted directly in terms of the load and displacement at the point of contact between the bar and the specimen. Dynamic analysis of the data is required as in the case of the instrumented impact test. Ruiz and Mines (1985) showed that in contrast to the noisy signal characteristic of the instrumented impact test, the strain signals observed in the Hopkinson bar are of much better quality. The last variation on this type of loading that we consider is due to Maigre and Rittel (1995). They considered a special type of specimen called the compact compression specimen; by sandwiching this specimen between two bars B and C as shown in Fig. 6.7c, and impacting bar B with a projectile A, they were able to generate high rate loading on the crack. Strain gages mounted on bars B and C were used to monitor the force and displacement at the points of contact between the specimen and the bars. These measurements were then used in a dynamic analysis to evaluate the dynamic stress intensity factor. They applied the method to the determination of the rate dependence of dynamic initiation toughness at rates in the range of $\dot{K}_{\mathrm{I}}^{\mathrm{dyn}}(t, v) = 10^5$ MPa$\sqrt{\mathrm{m}}$/s (Rittel and Maigre, 1995).

There are many other variations on this scheme of high strain rate testing; Klepaczko (1990) has provided a lengthy review of the use of Hopkinson pressure bars to dynamic fracture testing.

6.5 Explosives

Lead azide and pentaerythritol tetranitate (PETN) explosives have been used to apply high strain rate loading in the examination of dynamic fracture. However, issues related to the safe handling and operation of these explosives make it so difficult to use that this method of loading is not commonly employed. Furthermore, since the loading rates obtained are comparable to that obtained with moderate speed impact of projectiles, $\sim \dot{K}_I^{dyn}(t, v) = 10^5$ MPa$\sqrt{\text{m}}$/s, there are no significant advantages in using explosives. Some examples of dynamic fracture under explosive loading may be found in the work of Dally and Barker (1988) and Shukla and Rossmanith (1995); here we describe the experiment of Dally and Barker. In order to determine the dynamic crack initiation toughness of A533-B reactor grade steel, Dally and Barker developed an explosive loading scheme shown in Fig. 6.9. A long bar, 25–50 mm wide and 400–500 mm long, was cut into the dog-bone shape shown in the figure. A fatigue precrack was introduced at one edge in the middle of the bar. By detonating four explosive charges at the ends of the dog bone, a tensile stress pulse was sent towards the crack from both ends of the specimen. The resulting interaction at the crack tip generated a high rate loading at the crack tip; crack initiation was triggered within 10 μs; strain gages were used to determine the strain evolution near the crack tip; the time variation of the strain is shown in Fig. 8.16 and the results are discussed further in Chapter 8. The strain rate obtained in this experiment is $\dot{K}_I^{dyn}(t, v) = 8 \times 10^6$ MPa$\sqrt{\text{m}}$/s.

6.6 Electromagnetic Loading

The electromagnetic loading scheme is based on electromagnetic interaction between two current carrying conductors (Ravi-Chandar and Knauss, 1982). The specimen is made

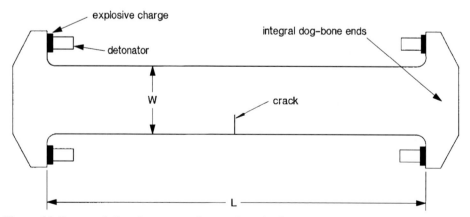

Figure 6.9 Four explosive charges are detonated at the four tabs on the dog-bone specimen to generate two tensile pulses that travel towards the crack from both sides of the specimen. (Reproduced from Dally and Barker, 1988.)

of a large plate (500 mm × 300 mm) of thickness 4.76 mm; a crack is introduced parallel to the long side by machining a 3-mm-wide slit down the middle. A natural crack tip is introduced by wedging a razor blade inside the slit and applying a small impact force. A flat copper strip, 4.76 mm × 1.2 mm thick is folded back on itself and the space between the two layers is filled with a mylar insulating strip. This assembly is then introduced into the machined slit as indicated in the schematic diagram in Fig. 5.9a. When a current flows through the copper loop, each leg generates a magnetic field surrounding it, with the magnetic field oriented normal to the current vector. The current vector in each leg interacts with the magnetic field of the other leg to produce an electromagnetic repulsion that forces the conductors apart. Since the two legs of the copper strip are confined in the slot of the machined crack, they do not move apart, but simply press upon the top and bottom surfaces of the crack with a uniform pressure. The magnitude of the pressure loading may be estimated easily from electromagnetic theory; for two idealized conductors of width b, carrying a current $i(t)$ the pressure of repulsion, in the limit where the separation between the two conductors is negligibly small in comparison to the conductor width b, is given by

$$p(t) = \frac{1}{2} \mu_0 \left(\frac{i(t)}{b} \right)^2 \tag{6.7}$$

where $\mu_0 = 4\pi \times 10^{-7}$ Wb/A m is the permeability constant. The current in the copper strip is generated by a discharge from a capacitor bank. The time history of the current which dictates the magnitude and duration of the pressure applied on the crack surface may be controlled by suitable choice of capacitors and inductors that form the pulse shaping circuit; Ravi-Chandar and Knauss (1982) generated a nearly trapezoidal pulse, with a rise to the peak amplitude in about 25 μs, and a total duration of about 150 μs. For typical values of current used in the experiments, the crack surface pressures were in the range of 1–20 MPa. For the large specimen, this loading configuration is equivalent to

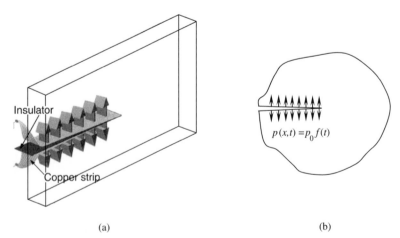

(a)   (b)

Figure 6.10 Principle of the electromagnetic loading scheme. (Reproduced from RaviChandar and Knauss, 1982.)

the crack deflection angle is given by

$$\gamma = -2\frac{K_{II}}{K_I^0} = -2C_{II}(\alpha, v)\cos(kx\cos\alpha + ky\sin\alpha - \omega x/v) \tag{7.5}$$

Therefore, the crack path oscillation will mirror this variation of the perturbation of the mode II stress intensity factor—a periodic modulation of the crack path is created as indicated schematically in Fig. 7.3. Noting that the amplitudes of the crack oscillations are likely to be very small, the second term in the argument of the sine function can be neglected. Then, the wave length of the ripple pattern observed on the crack surface is

$$\lambda_r = \frac{2\pi}{k\left|\cos\alpha - \dfrac{C}{v}\right|} \tag{7.6}$$

The ripple marks can be analyzed post-mortem to determine λ_r and used to determine the crack speed. Note that a longitudinal or a shear wave can be used in modulating the crack path; clearly, better spatial resolution is obtained with a shear wave. For a stress wave impinging at an angle $\alpha = \pi/2$, the crack speed is given by

$$v = \lambda_r f \tag{7.7}$$

where f is the frequency in Hz of the incident wave. A micrograph of the crack surface modulations observed in reflected light microscopy is shown in Fig. 7.4. The accuracy of the spatial measurement can be very high since one relies on a post-mortem examination

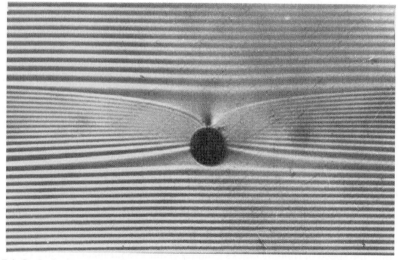

Figure 7.4 Optical micrograph of an ultrasonically modulated fracture surface; the dark region is a capillary tube of diameter 0.06 mm in the direction perpendicular to the crack surface. Crack growth is from bottom to top at a speed of about 150 m/s below the tube and ~ 20 m/s above the tube. Fringes are produced by the undulations in the fracture surface, when illuminated obliquely. (Reproduced from Kerkhof, 1973.)

of the fracture surface modulation at high magnifications; micron resolution in crack position identification is possible. The temporal resolution is dictated by the ultrasonic modulation frequency—1 MHz in Kerhkof's experiments. With typical crack speeds on the order of 10^3 m/s and frequency in the 1 MHz range, the wave length of the ripple pattern can be estimated to be 1 mm. At higher crack speeds the spacing increases and reduces the spatial resolution that may be obtained. At lower speeds, the ripple spacing decreases quickly and at a speed of about 0.02 m/s, one runs into limits of transmitting the stress wave into the material and observing the ripple spacing due to optical diffraction limit. One major limitation is the large attenuation of high-frequency waves as they propagate through short distances. Through suitable choice of the ultrasonic transducer, crack speeds in the range of 0.02–2000 m/s could be measured by this technique called *stress wave fractography*; a recent review of this technique can be found in Richter and Kerkhof (1994). Field (1971) and his colleagues have demonstrated the application of the ultrasonic modulation to a number of amorphous and crystalline materials. Kerkhof (1973) made careful measurements of the limiting speed in inorganic glasses by systematically varying the composition and showed that the composition had a significant influence on the terminal crack speed.

7.3 Electrical Resistance Methods

The *electrical resistance methods* in different manifestations have been used to determine the crack speed: *resistive grid methods* and *potential drop methods*. In the grid method a number of electrical wires are laid across the path of the crack. As the crack propagates, it severs the wires sequentially and provides an electrical signal which can then be used to determine the crack position and speed with time (Dulaney and Brace, 1960; Cotterell, 1965, 1968; Anthony et al., 1970; Paxson and Lucas, 1973). Commercial suppliers of strain gages now provide such grids for crack speed measurements; these grids can be incorporated into standard strain gage bridge circuits to provide the history of wire breakage and hence the crack position as a function of time. While very good estimates of the crack speed can be obtained from such grid techniques, the discrete nature of the grids dictate that the sampling rate of the crack speed will typically be much lower than that obtained using other methods such as ultrasonic modulation. However, if a thin conducting film is used instead of a grid, higher temporal resolution can be obtained; this is the basis of the potential drop technique. The resistance of the conducting layer changes as the crack length increases; it is also influenced by the film thickness and the crack opening. If this resistance change is measured, it can be related to the crack length and speed, while the other influences are negligible or minimized. Carlsson et al., 1973, demonstrated this application to the measurement of crack speed in PMMA; they used a voltage divider circuit to measure the resistance change. Many other investigators have used this method to determine the speed of running cracks (Stalder et al., 1983; Fineberg et al., 1991, 1992). Commercial versions of this technique—such as the KrakGage™—are now available; it is also quite easily accomplished in the laboratory with thin film coating methods. The discussion below follows the work of Hauch and Marder (1998) who used a Wheatstone bridge circuit for the measurement of resistance changes. The samples of PMMA,

Homalite, and glass were coated with a 20—50 nm thick coating of aluminum by evaporation in a vacuum of 10^{-6} Torr. The coating process naturally results in a spatial variation of the thickness, with a greater thickness at the center of the plate. Hausch and Marder describe how one can eliminate the thickness dependence of the resistance by calibration of each coating. Since there is no frequency dependence of the resistance, a calibration of the variation of resistance with the crack length may be performed with stationary cracks and then used to interpret the dynamic data. Crack opening displacement influences the measurements significantly; Carlsson et al. (1973) indicate that by choosing a proper frequency of the alternating current, the influence of crack opening may be decreased; they used a frequency of 3 MHz. On the other hand, Hauch and Marder used a direct current for measuring the resistance and considered the possibility of electrical discharge across an open crack; since the discharge results in clearly identifiable signals, they can be accounted for in the data analysis. See Hauch and Marder (1998) for a complete discussion of the factors that influence the potential drop technique. The variation of resistance with crack length is nearly linear and hence can be expressed as

$$R_{sp}(a) = R_0 + \frac{dR_{sp}}{da} da \tag{7.8}$$

R_0 and dR_{sp}/da can be determined from a calibration experiment or from a solution of Laplace's equation for the potential. Note that R_0 depends on the contact resistance and is not measured with great accuracy. However, this does not pose a problem since the initial resistance can be measured and subtracted. The instrumentation required for the measurement of this resistance is quite simple. The leads from the conducting surfaces across the crack line are connected to a Wheatstone bridge circuit. The resistance can be expressed in the following form

$$R_{sp}(a) = R_b \left[\frac{V_{bat}R_1 + V_{out}(a)(R_1 + R_a)}{V_{bat}R_a - V_{out}(a)(R_1 + R_a)} \right] \tag{7.9}$$

where V_{bat} is the input voltage derived from the battery, V_{out} is the measured output voltage, R_1, R_a, and R_b are the resistors used in the Wheatstone bridge circuit. From a measurement of V_{out} the crack length can be found through the calibration in Eq. 7.9. The change in output voltage with time can also be expressed in terms of the crack speed:

$$v = \frac{da}{dt} = \frac{(R_b + R_{sp})^2 V_{bat}}{R_b \dfrac{dR_{sp}}{da}} \frac{dV_{out}}{dt} \tag{7.10}$$

In order to increase the sensitivity of the method, the differentiation of the output voltage is accomplished through an analog circuit as indicated in Fig. 7.5. Both V_{out} and the analog differentiated dV_{out}/dt should be recorded with high accuracy and sampling rate to obtain the crack position and speed. According to Fineberg et al. (1992), this method provides for an evaluation of the crack position to within 200 μm and the crack speed to within 10 m/s. A sampling rate 10 M samples per second can be achieved—thus resulting in a very high temporal resolution of the measurements.

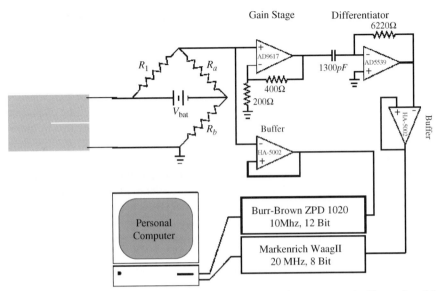

Figure 7.5 Potential drop technique for crack position and speed measurements. (Reproduced from Hauch and Marder, 1998.)

However, there remain some concerns still unaddressed. Crack tip process zone size is typically on the order of about 100 μm particularly at high load levels that tend to generate fast cracks. When the resistance is sampled at a rate of 10 M samples per second, a crack moving at 500 m/s moves about 50 μm between sampling intervals which is only about half of the process zone size; if the crack position is sampled at small time intervals such that the overall crack extension is less than the size of the fracture process zone, sampling introduces a random error—some spatio-temporal averaging may be desirable.

7.4 High-Speed Photography

Perhaps the most important technique for crack speed measurement is the *high-speed camera*; this has been a very popular although expensive method for dynamic fracture investigations. One major advantage of high-speed photography is that the event is observed without imposing any preconceived models for the interpretation of the observations. Another advantage is that full-field optical methods of stress and deformation analysis such as photoelasticity, method of caustics, shearing interferometry, moiré interferometry, and other techniques can be used to augment the crack position data with additional information on the crack tip deformation and stress fields. While the earliest attempts produced a single image by using a single spark light source, advances in the technology of high-speed photography have enabled camera designs for obtaining multiple images with a high spatial and temporal resolution. Modern high-speed cameras are capable of obtaining high-resolution images at time intervals on the order of 10 ns, with exposure times on the order of 2 ns. Field (1983) provides a good review of the techniques used in high-speed photography; modern high-speed cameras are available from many commercial vendors.

The first high-speed camera to obtain multiple images of crack growth was the Cranz-Schardin multiple-spark camera; in this camera system, multiple spark light sources are arranged in a square or rectangular array. The object or region of interest is imaged on a camera with multiple lenses, also arranged in the same square or rectangular array as the light source. Each spark then forms a discrete image through its matching lens. Riley and Dally (1969) provide a detailed description of the design, construction and operation of a Cranz-Schardin camera. Two major limitations exist in the Cranz-Schardin camera system: first, the multiple light sources cannot be aligned with the optical axis and hence a parallax error is introduced in the observations. Second, the duration of the spark sources is about 0.3 μs; this long exposure time results in an image smear, but for typical crack speeds of about 10^3 m/s, the image quality is quite acceptable since the smear is only about 300 μm. Large fields of view are possible in this arrangement; the system described by Riley and Dally has an 45.7 cm field of view. The timing between the sparks is controlled independently and hence the images can be obtained at different time intervals, focusing on times where higher data rates may be needed.

Rotating mirror cameras and rotating drum cameras provide an improvement over the Cranz-Schardin cameras in some respects, but not others. The operating principle in these cameras is to transport the image to a different location on the film either by moving the image with a rotating mirror or by moving the film by rotating the drum that holds the film and in some cases through a combination of the two mechanisms. Illumination from a spark light source and more recently from a pulsed laser is used to form discrete images. Parallax errors are eliminated in this arrangement since the light source and the imaging system are aligned along the optical axis. The number of images that can be obtained is increased significantly to more than a 100 frames. On the other hand, the size of the frames is usually quite small; this is dictated by the fact that during the inter-frame time interval the image has to be translated by a distance equal to its size. The speed with which this can be accomplished is limited by resonance phenomena in the turbines used to rotate the mirror or the drum. Typical framing rates that are possible with the rotating mirror or drum cameras is in the range of $10^3 - 10^6$ frames per second.

More recently, two types of electronic high-speed imaging cameras have been developed—the image converter cameras and CCD cameras. In the image converter cameras, the image falls on a photo-cathode which converts the image to a stream of electrons. These streams are steered to different parts of a phosphorescent screen by deflector plates, forming discrete images; the phosphorescent images are retained for many seconds and are then photographed on standard photographic film. Exposure time is controlled by the flash that illuminates the photo-cathode, but the framing rate is controlled by the speed of switching of the deflector plates. Therefore, very high framing rates may be obtained with these cameras. The main drawback is that the number of frames that can be obtained is usually quite small—about 4–12. The spatial resolution is dictated by the phosphorescent screen and is inferior to that achieved with rotating mirror cameras. Multiple CCD cameras have also been introduced recently; these are not too different from the Cranz-Schardin cameras in that arrays of CCDs are used to obtain multiple images.

However, the image is made to fall on the different sensors by using beam splitters thereby eliminating parallax errors. The CCDs are exposed at high speeds but are then read off-line at slower speeds. As in the image converters, while high speeds—on the order of 10^8 frames per second—are possible, the number of frames and the image resolution are limited. Typically about 8–16 frames are available and the image resolution is about 1300×1000 pixels over the image. New hybrid rotating mirror cameras with CCDs for image capture have also been developed with the capability to obtain large number of frames at extremely high framing rates.

Crack speed measurements provide two important conclusions that are quite independent of the details related to the resolution of the different crack speed measuring techniques. First, cracks attain a limiting speed of propagation that is significantly lower than the limit of the Rayleigh wave speed set by the continuum energy balance argument (see Section 5.1). Second, the limiting wave speed is not a fixed fraction of the characteristic wave speeds in the material, ranging anywhere from 0.33 C_R to 0.66 C_R (see Section 11.1 for further details). These conclusions point to the fact that while *a limiting speed is set by the continuum wave propagation theory, inherent material processes that govern fracture dictate a significantly lower limit*. In fact, Schardin (1959) suggested that the limiting crack speed be considered a new physical constant, perhaps related to other physical parameters that govern the fracture process. Along these lines, Kerkhof (1973) made an interesting observation: if the material ahead of the crack tip is breaking apart rapidly, then the relevant physical quantity is not the modulus, but the surface tension. In other words, if elastic energy propagation is characterized by the elastic modulus, then the surface energy must play an equivalent role in crack propagation (surface energy propagation?). Therefore, he suggested that the limiting speed must be proportional to $\sqrt{T/\rho}$ where T is the surface tension and ρ the density. However, since it is difficult to estimate the surface tension in solids, Kerkhof used the hardness σ_H in place of T and showed that the limiting speed in many inorganic glasses of different compositions can be related to $\sqrt{\sigma_H/\rho}$. It must be noted that the motivation for this correlation is purely phenomenological, but the correlation does exist. More recently, Gao (1996) has described a nonlinear continuum model that is essentially similar to the Kerkhof idea; σ_H is equivalent to the maximum cohesive stress σ_{max} in the Gao model. We will return to this issue in Chapters 11 and 12.

Chapter 8

Crack Tip Stress and Deformation Field Measurement

We now provide a discussion of various methods that have been used for the evaluation of the crack tip stress, strain or displacement fields. These include optical methods such as photoelasticity, moiré interferometry, the method of caustics and coherent gradient sensing developed specially for applications in fracture mechanics and strain gage methods.

8.1 Jones Calculus

Propagation of a polarized light beam through optical components that exhibit refractive index changes can be analyzed with great ease through the matrix approach introduced by Jones (1941); *Jones calculus* is summarized here in order to facilitate discussions of photoelasticity and lateral shearing interferometry. Detailed account of the method can be found in the monograph by Srinath (1983). A polarized light beam can be represented by the amplitude of its electric field, decomposed along two directions orthogonal to the propagation direction; let us take the propagation direction to be x_3 and let (x_1, x_2, x_3) form a right-handed coordinate frame. The electric vector components are then written in a vector form, called the *Jones vector*

$$
\mathbf{E} = \begin{bmatrix} E_1 e^{i(\omega t - \varphi_1)} \\ E_2 e^{i(\omega t - \varphi_2)} \end{bmatrix}
\tag{8.1}
$$

where E_1 and E_2 are the amplitudes resolved along the x_1 and x_2 directions, $\omega = 2\pi c/\lambda$, c the speed of light, λ the wave length of light, and φ_1 and φ_2 are the absolute phase angles of the two components; in general, this Jones vector represents an elliptically polarized light beam. In evaluating the influence of any optical element on light propagation, we will be concerned with the relative phase difference introduced between these two components of the light vector and hence for convenience, we can set one of the phase changes to unity and carry the relative phase difference in the other component. Let the relative phase angle $\Delta s \equiv \varphi_1 - \varphi_2$. Each optical element in the path of the light beam is characterized by two parameters—the orientation, α, of its fast principal optical axis relative to the global x_1 direction and the amount of relative phase change Δs that the optical element introduces;

with these two parameters, the Jones vector of the light incident on the optical element can be transformed to determine the Jones vector of the light leaving the optical element. The transformation matrix is called the *Jones matrix* and is easily calculated. We will illustrate this for a general optical element and then apply it to particular cases. The transformation is performed in three steps: first, the incoming wave $\mathbf{E}_{in}$ is resolved along the principal directions of the optical element:

$$\mathbf{E}' = \begin{bmatrix} \cos\alpha & \sin\alpha \\ -\sin\alpha & \cos\alpha \end{bmatrix} \mathbf{E}_{in} \tag{8.2}$$

Second, the relative phase change (a retardation along the slow axis relative to the fast axis) is introduced along the slow principal direction

$$\mathbf{E}'' = \begin{bmatrix} 1 & 0 \\ 0 & e^{-i\Delta s} \end{bmatrix} \mathbf{E}' \tag{8.3}$$

where $i = \sqrt{-1}$ and Δs the relative phase difference between the principal (the fast and slow) components of the light vector. Finally, the principal components are recombined along the global x_1, and x_2 directions.

$$\mathbf{E}_{out} = \begin{bmatrix} \cos\alpha & -\sin\alpha \\ \sin\alpha & \cos\alpha \end{bmatrix} \mathbf{E}'' \tag{8.4}$$

Combining these operations, the relationship between the input Jones vector $\mathbf{E}_{in}$ and the output Jones vector $\mathbf{E}_{out}$ for this optical element is given by

$$\mathbf{E}_{out} = \mathbf{J}(\Delta s, \alpha)\mathbf{E}_{in} \tag{8.5}$$

where

$$\mathbf{J}(\Delta s, \alpha) = \begin{bmatrix} \cos^2\alpha + e^{-i\Delta s}\sin^2\alpha & \sin\alpha \cos\alpha(1 - e^{-i\Delta s}) \\ \sin\alpha \cos\alpha(1 - e^{-i\Delta s}) & \sin^2\alpha + e^{-i\Delta s}\cos^2\alpha \end{bmatrix} \tag{8.6}$$

is the Jones matrix for the optical element. If the light beam passes through a collection of n optical elements each with Jones matrix $\mathbf{J}_k$, then the output of each element is the input for the succeeding element; hence, the output electric vector is given by

$$\mathbf{E}_{out} = \mathbf{J}_n(\Delta s_n, \alpha_n)\mathbf{J}_{n-1}(\Delta s_{n-1}, \alpha_{n-1})\cdots\mathbf{J}_2(\Delta s_2, \alpha_2)\mathbf{J}_1(\Delta s_1, \alpha_1)\mathbf{E}_{in} \tag{8.7}$$

For future reference, we list below the Jones matrices of common optical elements: a linear polarizer, a quarter wave plate, and a half-wave plate. A linear polarizer allows only one plane of polarization; therefore, the relative phase change can be taken to be infinite in the other component; let the orientation be at an angle α with respect to the global direction. Then the Jones matrix is given by

$$\mathbf{J}_{linear}(\infty, \alpha) = \begin{bmatrix} \cos^2\alpha & \sin\alpha \cos\alpha \\ \sin\alpha \cos\alpha & \sin^2\alpha \end{bmatrix} \tag{8.8}$$

For a quarter wave plate, $\Delta s = \pi/2$; it is usually placed at an angle of $\alpha = \pm\pi/4$ with respect to the global directions; therefore the Jones matrix is given by

$$\mathbf{J}_{\lambda/4}\left(\frac{\pi}{2}, \pm\frac{\pi}{4}\right) = \begin{bmatrix} \dfrac{1-i}{2} & \pm\dfrac{1+i}{2} \\ \pm\dfrac{1+i}{2} & \dfrac{1-i}{2} \end{bmatrix} \tag{8.9}$$

An isotropic phase retarder changes the phase in each component by the same amount; the corresponding Jones matrix is

$$\mathbf{J}_{\text{iso}}(\varphi, \alpha) = \begin{bmatrix} \exp(-i\varphi) & 0 \\ 0 & \exp(-i\varphi) \end{bmatrix} \tag{8.10}$$

With the Jones calculus of polarized light, we can examine light propagation in a stressed specimen in specific optical configurations.

8.2 Photoelasticity

Brewster (1814, 1815) found that in some materials the index of refraction was affected by the application of pressure; upon removal of the external load, the index of refraction returned to its original value. The basis of the method of photoelasticity is this temporary, stress-induced birefringence exhibited by many polymers; for example polymethylmetha-crylate, Homalite-100 and polycarbonate are commonly used in dynamic fracture investigations because of their birefringent properties. Thus, a stressed specimen becomes a phase retarder (with spatially varying retardation induced by the spatial variation of the stress field); the fast and slow principal optical directions coincide with the principal stress directions. The index of refraction at a point on a stressed specimen depends on the stress state at that point; hence a polarized light beam traveling through a stressed specimen will experience a relative phase retardation in the components of the light resolved along the local principal stress directions. This stress-induced birefringence can be exploited to reveal the stress field in the specimen. In this section, we first describe the phenomenon of temporary birefringence, then evaluate the Jones matrix for a stressed specimen and finally describe the circular polariscope that is commonly used for revealing the shear stress distribution in the specimen.

Maxwell expressed the dependence of the index of refraction on the principal stresses in the following form

$$\begin{aligned} n_1 - n_0 &= c_1\sigma_1 + c_2(\sigma_2 + \sigma_3) \\ n_2 - n_0 &= c_1\sigma_2 + c_2(\sigma_3 + \sigma_1) \\ n_3 - n_0 &= c_1\sigma_3 + c_2(\sigma_1 + \sigma_2) \end{aligned} \tag{8.11}$$

where n_0 is the isotropic, unstressed index of refraction, c_1 and c_2 are the direct and transverse stress optic coefficients (in units of m^2/N, sometimes labeled *brewsters*), σ_i the principal stress components at any point and n_i the refractive indices in the principal stress directions at that point. For the plane stress conditions that are typical of the thin plates

Table 8.1 Material stress fringe value f_σ for selected polymers (λ = 514 nm)

Material	f_σ (kN/m)	Source
Homalite-100	22.2	Dally and Riley (1978)
Polycarbonate	6.6	Dally and Riley (1978)
Polymethylmethacrylate	129	Kalthoff (1987)

used in most fracture experiments, σ_3 can be set to zero; furthermore, for light rays propagating in the x_3 direction, only the refractive indices n_1 and n_2 are of interest. Hence, a polarized light beam that travels through a point on a stressed specimen will be decomposed into two components along the principal directions and these two components travel with different speeds; thus the light components that emerge from the specimen will have a relative angular phase difference given by

$$\Delta s(x_1, x_2) = \frac{2\pi h}{\lambda}(n_1 - n_2) = \frac{2\pi h C}{\lambda}(\sigma_1 - \sigma_2) = \frac{2\pi h}{f_\sigma}(\sigma_1 - \sigma_2) \tag{8.12}$$

where $C = (c_1 - c_2)$ is the relative stress optic coefficient, h the thickness of the specimen through which the light travels, and $f_\sigma = \lambda/C$ is called the *material stress fringe value*. Methods for the determination of f_σ are discussed in books dealing with photoelasticity (see for example, Dally and Riley, 1978). Typical values of f_σ for different materials are listed in Table 8.1. The spatial dependence of the phase difference arises from the spatial dependence of the stress field. Thus, the effect of a stressed birefringent specimen on light propagation is simply that a phase retardation is introduced between components resolved along the local principal directions. Let $\beta(x_1, x_2)$ denote the orientation of the local maximum principal direction with respect to the global x_1 direction at any point in the specimen. The effect of the stressed specimen on the propagation of a polarized light beam at the point (x_1, x_2) may be represented using a Jones matrix:

$$\mathbf{J}_b(\Delta s, \beta) = \begin{bmatrix} \cos^2\beta + e^{-i\Delta s}\sin^2\beta & \sin\beta \cos\beta(1 - e^{-i\Delta s}) \\ \sin\beta \cos\beta(1 - e^{-i\Delta s}) & \sin^2\beta + e^{-i\Delta s}\cos^2\beta \end{bmatrix} \tag{8.13}$$

In order to exhibit the phase difference in a visual form, the stressed specimen is introduced into a circular polarizer; the arrangement of a circular polarizer is shown schematically in Fig. 8.1. Consider a light beam polarized along the global x_2 direction. The incoming light is described by the Jones vector:

$$\mathbf{E}_{\text{in}} = \begin{bmatrix} 0 \\ k e^{i\omega t} \end{bmatrix} \tag{8.14}$$

It is then made to pass through a quarter wave plate that introduces a relative retardation of one quarter wavelength between the two components of the incident light vector. The fast axis of the quarter wave plate is oriented at an angle $\pi/4$ with respect to the global x_1 direction; the Jones matrix of the quarter wave plate is then evaluated from Eq. 8.6 with $\Delta s = \pi/2$, $\alpha = \pi/4$ and is given in Eq. 8.9. The light vector exiting the quarter wave plate

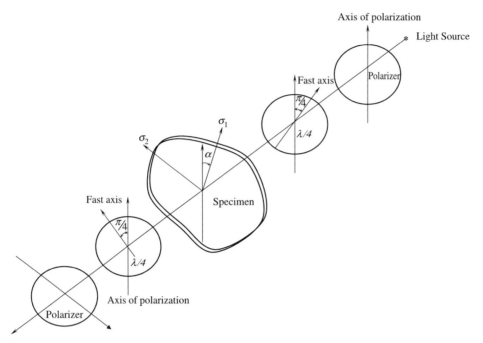

Figure 8.1 Optical arrangement of a circular polariscope in a dark field configuration.

is easily determined through Jones calculus:

$$\mathbf{E}' = \mathbf{J}_{\lambda/4}\left(\frac{\pi}{2}, \frac{\pi}{4}\right)\mathbf{E}_{\text{in}} = k e^{i\omega t}\begin{bmatrix} \dfrac{1+i}{2} \\ \dfrac{1-i}{2} \end{bmatrix} \tag{8.15}$$

The light beam exiting the quarter wave plate is a circularly polarized light; this beam is then transmitted through the stressed specimen where it suffers a phase retardation in one component relative to the other through the stress-induced birefringence discussed above. The light vector exiting the specimen accumulates a phase difference given in Eq. 8.12 and is represented by

$$\mathbf{E}'' = \mathbf{J}_b(\Delta s, \beta)\mathbf{J}_{\lambda/4}\left(\frac{\pi}{2}, \frac{\pi}{4}\right)\mathbf{E}_{\text{in}} \tag{8.16}$$

The light beam exiting the specimen is then made to pass through a second quarter wave plate oriented at $-\pi/4$ with respect to the global x_1 axis and then through a second polarizer. The second polarizer may be oriented either along the x_1 direction (called the *dark field arrangement*) or along the x_2 direction (called the *bright field arrangement*). The light beam exiting the second polarizer in the dark field arrangement can then be written as

$$\mathbf{E}_{\text{out}} = \mathbf{J}_{\text{linear}}(\infty, 0)\mathbf{J}_{\lambda/4}\left(\frac{\pi}{2}, -\frac{\pi}{4}\right)\mathbf{J}_b(\Delta s, \beta)\mathbf{J}_{\lambda/4}\left(\frac{\pi}{2}, \frac{\pi}{4}\right)\mathbf{E}_{\text{in}} \tag{8.17}$$

Substituting for the input from Eq. 8.14 and for the appropriate Jones vectors from Eqs. 8.6, 8.8 and 8.9, the output can be shown to be

$$\mathbf{E}_{\text{out}} = \frac{ke^{i\omega t}}{2} \mathbf{J}_{\text{linear}}(\infty, 0) \begin{bmatrix} e^{-i2\beta}(1 - e^{-i\Delta s}) \\ -i(1 + e^{-i\Delta s}) \end{bmatrix} \tag{8.18}$$

The linear polarizer placed along x_1 direction allows only the x_1 component of the light to pass through and therefore the output electric vector is given by

$$\mathbf{E}_{\text{out}} = \frac{ke^{i\omega t}}{2} \begin{bmatrix} e^{-i2\beta}(1 - e^{-i\Delta s}) \\ 0 \end{bmatrix} \tag{8.19}$$

The intensity of this light beam is the time average of $\mathbf{E}_{\text{out}}^2$, averaged over a time significantly longer than the period:

$$I(x_1, x_2) = \langle \mathbf{E}_{\text{out}}^2 \rangle = \frac{k^2}{2} \sin^2 \left[\frac{\Delta s(x_1, x_2)}{2} \right] \tag{8.20}$$

Thus, the spatial variation of the light intensity is governed only by the phase retardation introduced by the stressed specimen; in this optical arrangement, the orientation of the principal directions $\beta(x_1, x_2)$ does not influence the intensity. Introducing the phase difference from Eq. 8.12, bright fringes corresponding to maximum light intensity are lines in the $x_1 - x_2$ plane along which

$$(\sigma_1 - \sigma_2) = \frac{Nf_\sigma}{h}, \text{ with } N = 0, \pm 1, \pm 2, \dots \tag{8.21}$$

where N is called the fringe order. Therefore, placing the specimen between crossed circular polarizers reveals lines of constant intensity that are contours of constant in-plane shear stress; these lines are called *isochromatic fringes*. This is the basis of all photoelasticity; in applications to dynamic problems, the isochromatic fringe patterns are captured with a high-speed camera at short time intervals to provide a time history of the evolution of the shear stresses in the specimen. In applications to fracture mechanics, it is assumed that the crack tip asymptotic field is applicable in the vicinity of the crack tip and the parameters of the asymptotic field are extracted by fitting the observed isochromatic fringes to theoretically predicted patterns in a least-squared error process. This procedure is discussed in Section 8.2.1.

8.2.1 Evaluation of the Dynamic Stress Intensity Factor using Photoelasticity

Assuming that the fringe pattern formation is governed by the asymptotic stress field near the crack tip, the geometry of the fringe pattern can be expressed as follows

$$\frac{Nf_\sigma}{h} = (\sigma_1 - \sigma_2) = g(r, \theta; K_{\text{I}}^{\text{dyn}}, K_{\text{II}}^{\text{dyn}}, \sigma_{\text{ox}}) \tag{8.22}$$

where N is the fringe order, f_σ the fringe sensitivity, h the specimen thickness, and σ_1, σ_2 are the principal stress components. $g(r, \theta; K_{\text{I}}^{\text{dyn}}, K_{\text{II}}^{\text{dyn}}, \sigma_{\text{ox}})$ is determined from the linear

the change in thickness due to Poisson contraction of the specimen. This approximation can then be used to determine the ray paths. The position vector of the rays in the screen plane can be expressed as

$$\mathbf{R} = \mathbf{r} - \mathbf{w} = \mathbf{r} - z_0 \ \text{grad} \ \delta s(r, \theta) \tag{8.32}$$

where z_0 is the distance from the specimen midplane to the screen plane. Eq. 8.32 represents the transformation relation for mapping points $\mathbf{r} = x_1 \mathbf{e}_1 + x_2 \mathbf{e}_2$ in the specimen plane to points $\mathbf{R} = X_1 \mathbf{e}_1 + X_2 \mathbf{e}_2$ in the screen plane. $(\mathbf{e}_1, \mathbf{e}_2)$ are the unit vectors in the (x_1, x_2), directions. If the Jacobian determinant of this transformation is zero, then the mapping is singular; this is the condition for the formation of a caustic curve on the screen. Before examining the details of the mapping, it is useful to examine the mapping qualitatively.

As can be seen from the illustration in Fig. 8.4, far away from the crack tip, the light rays pass through the transparent specimen and maintain their parallel propagation; the influence of the stress field on the wavefront is small enough to be neglected. On the other hand, in the region near the crack tip, where the specimen exhibits a concave surface due to the Poisson contraction, the light rays deviate significantly from parallelism. As a result, a dark region called the *shadow spot* forms on the screen at z_0 where there are no light rays at all. This shadow region is surrounded by a bright curve, called the *caustic curve*. The line on the specimen plane whose image is the caustic curve on the specimen is called the *initial curve*. Light rays from outside the initial curve fall outside the caustic, rays from inside the initial curve fall on or outside the caustic curve and rays from the initial curve fall on the caustic curve. Hence the caustic curve is a bright curve that surrounds the dark region. Therefore, the equation for the initial curve is obtained by setting the Jacobian determinant of Eq. 8.32 to zero and the caustic curve is obtained by evaluating the mapping in Eq. 8.32 on the initial curve. The details of these calculations are shown below.

Using Eqs. 8.28 and 8.31 in Eq. 8.32 and restricting attention to mode I loading condition results in the following mapping equation

$$X_1 = r_d \left\{ \cos\theta_d + \frac{2}{3\alpha_d} \left(\frac{r_0}{r_d} \right)^{5/2} \cos\frac{3\theta_d}{2} \right\}$$

$$X_2 = \frac{r_d}{\alpha_d} \left\{ \sin\theta_d + \frac{2\alpha_d}{3} \left(\frac{r_0}{r_d} \right)^{5/2} \sin\frac{3\theta_d}{2} \right\} \tag{8.33}$$

where

$$r_0 = \frac{3hc_t z_0 K_I}{2\sqrt{2\pi}F(v)} \quad \text{and} \quad F(v) = \frac{2(1 + \alpha_s^2)(\alpha_d^2 - \alpha_s^2)}{4\alpha_d \alpha_s - (1 + \alpha_s^2)^2} \tag{8.34}$$

α_d and α_s are defined in Appendix A. A simulation based on the mapping equations in Eq. 8.33 is shown in Fig. 8.6a. The 'shadow spot' and the 'caustic curve' are clearly seen in this simulation. We can now evaluate the Jacobian determinant of the optical mapping

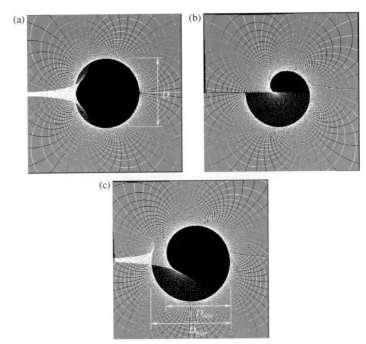

Figure 8.6 Simulated bitmap image of the transformation in Eq. 8.32. The dark region is the 'shadow spot' and is surrounded by the bright 'caustic' curve. The field of view represents a square, 20 mm to a side. (a) $K_I = 1$ MPa m$^{1/2}$, $K_{II} = 0$, (b) $K_I = 0$, $K_{II} = 1$ MPa m$^{1/2}$, (c) $K_I = 1$ MPa m$^{1/2}$, $K_{II} = 1$ MPa m$^{1/2}$.

in Eq. 8.33; this yields the equation for the initial curve:

$$\left(\frac{r_d}{r_0}\right)^5 - 1 + \left(\frac{r_d}{r_0}\right)^{5/2}(\alpha_d^2 - 1)\cos\frac{5\theta_d}{2} = 0 \tag{8.35}$$

The Jacobian determinant is zero at points (x_1, x_2) or (r_d, θ_d) that satisfy Eq. 8.35. Kalthoff (1987) and Rosakis (1980) showed that the last term in Eq. 8.35 is small when crack velocities are less than about 40% of the shear wave speed; since mode I cracks seldom move at speeds greater than this value, they suggested that this term could be neglected. Ignoring the last term, the Jacobian determinant goes to zero when $r_d = r_0$; therefore, the initial curve is given by

$$r_d = r\sqrt{1 - (v\sin\theta/C_d)^2} = r_0 \tag{8.36}$$

This represents a circle in the distorted coordinates $(x_1, \alpha_d x_2)$. Substituting this result into Eq. 8.34 yields the equation of the caustic curve

$$
\begin{aligned}
X_1^c &= r_0\left\{\cos\theta_d + \frac{2}{3\alpha_d}\cos\frac{3\theta_d}{2}\right\} \\
X_2^c &= \frac{r_0}{\alpha_d}\left\{\sin\theta_d + \frac{2\alpha_d}{3}\sin\frac{3\theta_d}{2}\right\}
\end{aligned}
\tag{8.37}
$$

where r_0 is given in Eq. 8.35. Eqs. 8.36 indicate that the size of the caustic depends only on r_0, which in turn depends directly on the plate thickness h, the optical coefficient c_t, the distance between the specimen midplane and the screen z_0, the dynamic stress intensity factor K_I, and only weakly on the crack velocity.

Eqs. 8.37 are the equations of an epicycloid, with r_0 playing the role of a scale parameter. The maximum extent of the caustic curve in the x_2 direction, D (marked on Fig. 8.6 and usually referred to as the *transverse diameter of the caustic curve*) is taken to be the a measure of the size of the caustic curve. Then, from Eqs. 8.37, the following relationship can be obtained:

$$D = [X_2(\theta = -\pi/2) - X_2(\theta = \pi/2)] = 3.17 r_0$$

Inserting this in Eq. 8.35 and rearranging establishes the following equation for the determination of the dynamic stress intensity factor:

$$K_I = \frac{2\sqrt{2\pi}}{2c_t h z_0 F(v)} \left(\frac{D}{3.17}\right)^{3/2} \tag{8.38}$$

Thus, from measurements of the transverse diameter of the caustic curve, the dynamic stress intensity factor can be determined. Application of the method to quasi-static and dynamic problems has been demonstrated by a number of investigators. If these measurements are performed by obtaining high-speed photographs of the images in the screen plane, the variation of the dynamic stress intensity factor with time may be obtained. An example of the caustics observed in a dynamically propagating crack is shown in Fig. 8.7.

8.3.2 Caustic in Reflection

The analysis described above is for transmission of light through a transparent specimen; the analysis can be repeated for the formation of caustics in an opaque specimen by reflection. The following changes need to be taken into account. First, the screen plane is behind the specimen and therefore the screen plane is a virtual plane obtained by focusing the camera to this plane; the optical arrangement is shown in Fig. 8.4. Second, the stress optic effect is not relevant and only the thickness change contributes to the caustic formation. Therefore, in Eq. 8.31, the surface of constant phase is replaced by

$$\frac{\Delta s}{2\pi/\lambda} = \delta s(x_1, x_2) = 2u_3\left(x_1, x_2, -\frac{h}{2}\right) = 2\int_{-h/2}^{h/2} \varepsilon_{33}\, dx_3$$

$$= -\frac{h\nu}{E}(\sigma_{11} + \sigma_{22}) \equiv hc_r(\sigma_{11} + \sigma_{22}) \tag{8.39}$$

where $c_r = -\nu/E$. Replacing δs in Eq. 8.32 with the above expression, all equations corresponding to caustics by reflection may be obtained. In particular, the relationship between the transverse diameter and the dynamic stress intensity factor is still given by Eq. 8.38, by simply replacing c_t with c_r.

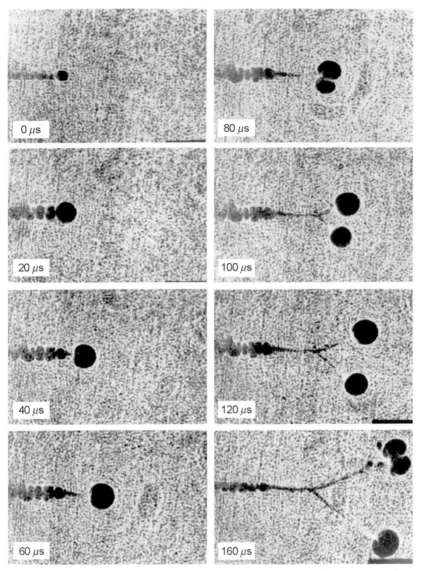

Figure 8.7 High-speed photographs of caustics at the tip of a dynamically propagating crack. (Reproduced from Ravi-Chandar and Knauss, 1984c.)

The analysis presented above corresponds to a crack growing at a constant speed v. Taking a limit as $v \to 0$ in Eq. 8.28, the derivation of the caustic curve appropriate for a stationary crack may be obtained. Note that the limit must be taken carefully because the function $f_{\alpha\beta}^1(\theta, v)$ becomes indeterminate in the limit as $v \to 0$. In this case, the relationship between the transverse diameter and the dynamic stress intensity factor is still given by Eq. 8.38, with $F(v \to 0)$ equal to 1.

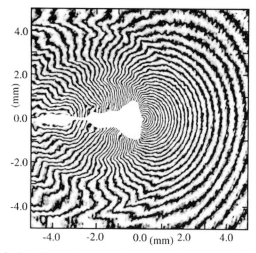

Figure 8.19 Fringes obtained in a Twyman–Green interferometer displaying the out-of-plane displacement field near a rapidly growing crack in a PMMA specimen. Post-processing was used to subtract out the initial fringe pattern. Note that a region near the crack tip is obscured due to aperture limitations and this region has been removed during image processing. Stress waves radiating from the crack tip damage zone can be seen as perturbations in the fringe pattern, especially behind the crack tip. The crack speed was 0.52 mm/μs, (about $0.522C_s$). (Reproduced from Pfaff et al., 1995.)

the singular term of the asymptotic stress field is given by:

$$u_\alpha(r, \theta) = \frac{\nu K_I^{\text{dyn}}(t, v)}{E\sqrt{2\pi}} \sqrt{r} g_\alpha^I(\theta; v)$$

$$u_3(r, \theta) = \frac{\nu K_I^{\text{dyn}}(t, v)}{E\sqrt{2\pi r}} f_{\alpha\alpha}^I(\theta; v)$$

(8.62)

where $g_\alpha^I(\theta; v)$ and $f_{\alpha\beta}^I(\theta; v)$ are known functions of θ and v given in Appendix A. The fringe patterns observed in the interferometry experiments are contours of constant displacement component: u_3 in the case of classical interferometry, and u_1 or u_2 in the case of in-plane moiré interferometry. An example of the fringes obtained in a Twyman–Green interferometer indicating the variation of the out-of-plane displacement field is shown in Fig. 8.19. The fringe patterns in Fig. 8.19 indicate that stress waves emanate from the crack tip at discrete times; these are identified by the perturbations in the fringe contours and are caused by the intermittency in the crack propagation process. Fitting the observed fringe patterns ahead of the crack to a prediction based on Eq. 8.62 using a least-square error method, the dynamic stress intensity factor can be determined, in much the same way as for the methods of photoelasticity and coherent gradient sensing. A major difficulty in using these methods arises from the fact that while the measurements of the displacement components are themselves very accurate, the application of the plane elastodynamic fields in the interpretation of the crack tip field parameters introduces significant errors, particularly when one approaches close to the crack tip. Furthermore the interferometric methods indeed pose a challenging task in dynamic situations and very few successful attempts have been reported in the literature.

Chapter 9

Dominance of the Asymptotic Field

The basic theory of elastodynamic fracture, the experimental methods used in generating dynamically growing cracks, the diagnostic tools for use in the evaluation of the crack position and crack tip stress field have been discussed in the preceding chapters. We now turn to a critical analysis of the theory through a comparison of the theory with experimental measurements. The limitations of the experimental methods in the determination of the dynamic stress intensity factor as well as the validity of characterizing the dynamic crack tip stress field through this parameter are discussed in this chapter. This is then followed by a discussion of the experimental determination of dynamic failure criteria in Chapter 10. The mechanisms that govern dynamic fracture and phenomenological models that have been developed in order to capture these mechanisms are presented in Chapters 11 and 12.

9.1 Stationary Cracks

The semi-infinite crack geometry with uniform pressure loading is amenable to experimental implementation with the use of an electromagnetic loading scheme (Ravi-Chandar and Knauss, 1982) described in Chapter 6. The specimen is made of a large Homalite-100 plate (500 mm × 300 mm) of thickness 4.76 mm; selected mechanical and optical properties of this material are given in Table B.1 in Appendix B. A crack is introduced parallel to the long side by machining a 3 mm wide slit down the middle. A natural crack tip is introduced by wedging a razor blade inside the slit and applying a small impact force. A flat copper strip, 4.76 mm × 1.2 mm thick is folded back on itself and the space between the two layers is filled with a mylar insulating strip. This assembly is then introduced into the machined slit as indicated in the schematic diagram in Fig. 6.9. Insertion of this assembly introduced a small static loading on the crack tip, which was evident in the experimental measurements as we shall discuss later. When a current flows through the copper loop, each leg generates a magnetic field surrounding it, with the magnetic field oriented normal to the current vector. The current vector in each leg interacts with the magnetic field of the other leg to produce an electromagnetic repulsion that forces the conductors apart. Since the two legs of the copper strip are confined in the slot of the machined crack, they press upon the top and bottom surfaces of the crack with a uniform pressure. The current in the copper strip is generated by a discharge from a capacitor bank. The magnitude of the pressure loading may be estimated easily from electromagnetic theory; the time history of the current, which dictates the magnitude and duration of the pressure applied on the crack surface, may be controlled by suitable choice of capacitors and inductors that form the pulse-shaping circuit. Ravi-Chandar and Knauss (1982) generated a nearly trapezoidal pulse, with a rise to the peak amplitude in about 25 μs and a total duration of about 150 μs. For typical values of current used in the experiments, the crack surface pressures were in the range of 1–20 MPa. For the large specimen, this loading configuration is equivalent to an infinite plate, with a pressurized semi-infinite crack for the duration of the current pulse, conforming to the boundary-initial value problems discussed in Chapter 4. The crack tip response to the loading was monitored to determine the time variation of the dynamic stress intensity factor. A high-speed camera capable of capturing images with a 15 ns exposure time and 5 μs time interval between frames was used; the optical method of caustics (described in Chapter 8) was used to determine the dynamic stress intensity factor. A typical sequence of high-speed photographs obtained in this experimental arrangement is shown in Fig. 8.7.

From a series of these experiments, Ravi-Chandar and Knauss (1982, 1984c) evaluated the dynamic stress intensity factor as a function of time and compared it with the theoretical estimates in Eq. 4.19. Two sets of results from their experiments are reproduced here. First, Ravi-Chandar and Knauss (1982) considered a partially loaded semi-infinite crack. The time history of the dynamic stress intensity factor shown in Fig. 9.1 corresponds to a pressure load distributed over a length l at a distance L behind the crack tip. The theoretical variation of the dynamic stress intensity factor was obtained using the superposition integral in Eq. 4.37. Clearly, the experimental measurements and theoretical predictions agree well within the experimental accuracy. The slight drop in the stress intensity factor at the arrival of the dilatational wave that is indicated in the theoretical prediction is not easily observed in the experiment due to limitation in the resolution. Kim (1985a,b) used the same loading apparatus, but a different optical diagnostic method for the evaluation of the dynamic stress intensity factor. Instead of photographing the caustics, he brought the light rays to a focus and filtered the focal volume; therefore, the only rays that pass through are the rays that are deviated by the crack tip deformation. Kim was then able to analyze the intensity of these rays collected at a photodetector in terms of the dynamic stress intensity factor. Due to the higher time resolution of the photodetector compared with the high-speed camera, he was able to obtain stress intensity factor measurements with a better time resolution; in his experiments, a length L near the crack tip was left without the pressure loading. His results also indicated a good agreement between the measured dynamic stress intensity factor and that calculated from Eq. 4.40, including the initial reduction in the stress intensity factor from the compressive dilatational wave.

In the second series of tests, a uniform pressure of magnitude in the range of 0.63–15.4 MPa was applied over the entire crack corresponding to a pressurized semi-infinite crack. Experimental measurements of the dynamic stress intensity factor corresponding to a stationary crack for four different pressure levels are shown in Fig. 9.2.

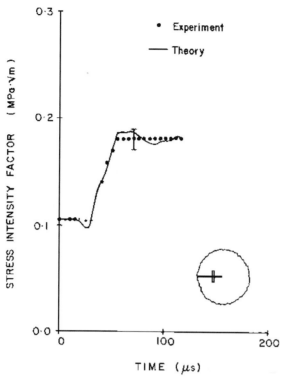

Figure 9.1 Comparison of the measured time history of the dynamic stress intensity factor for a semi-infinite crack in an unbounded medium with the theoretical prediction. (Reproduced from Ravi-Chandar and Knauss, 1982.)

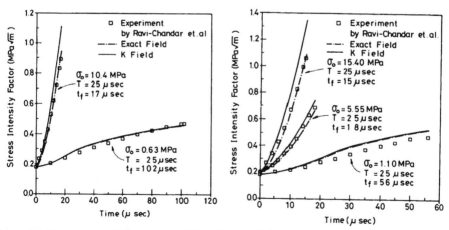

Figure 9.2 Comparison of the measured time history of the dynamic stress intensity factor for a semi-infinite crack in an unbounded medium with the theoretical prediction. (Reproduced from C.C.Ma, 1991.)

These measurements are compared to the theoretical prediction from Eq. 4.19. Clearly, very good agreement is demonstrated; note that in evaluating the theoretical predictions, a trapezoidal fit to the actual time history of the pressure loading was used. Ma (1991) made a further improvement in this by considering the exact stress field ahead of the crack indicated in Eq. 4.28 rather than by assuming that the singular term alone was sufficient to represent the stress field. In all these experiments, the dynamic stress intensity factor reached a sufficiently high value to cause crack growth to occur. The results corresponding to crack initiation and growth will be examined later.

The experiments described here are the only ones where a comparison of theoretical estimates of the dynamic stress intensity factor could be obtained directly; however, there are many other experimental investigations where the dynamic stress intensity factors have been evaluated through different diagnostic methods—such as photoelasticity, shearing interferometry or CGS, and strain gauges—and compared to numerical simulations (we will not delve into these here, but see for example the works of Kobayashi, Dally, Kalthoff, Shukla, Rosakis, Takahashi and others listed in the bibliography).

9.2 Propagating Cracks

The experiments described above also provided the means of evaluating the dynamic stress intensity factor history for dynamically propagating cracks. We examine the case of propagating cracks in this section. In this series of experiments, crack initiation occurred in the time interval: $15\mu s < \tau < 110\mu s$. Continued growth of the crack was observed to occur at a constant speed with acceleration to the final speed occurring within the time resolution of the measurement (less than 5 μs). Fig. 9.3 shows a comparison of the experimentally measured time variation of the dynamic stress intensity factor with the dynamic stress intensity factor estimated from the analysis described in Section 4.2. The experimentally observed variation of the crack position with time is also shown in Fig. 9.3. The crack surface pressure in this experiment was 1.10 MPa. Upon loading, the dynamic stress intensity factor increased gradually until 56 μs when the crack began to propagate. Crack extension was observed to occur at a constant speed of 240 m/s ($= 0.22C_R$), with a corresponding drop in the stress intensity factor. Beyond 150 μs, waves from the finite boundaries of the specimen arrived at the crack tip to load it further with a stress pulse; it is possible to analyze the results in this time range, but this is not of immediate interest. As pointed out earlier, the analysis provides dynamically admissible solutions; in order to pick the correct solution, the experimentally measured time of crack initiation and crack speed were used. Thus, imposing the experimental observations that the crack initiated at time τ and propagated with a speed v, the dynamic stress intensity factor was calculated; this expression is given in Eq. 4.77. The comparison between the theoretical estimate and the experimental measurement shown in Fig. 9.3 is remarkably good, confirming the validity of the elastodynamic stress analysis of cracks under the conditions presented here.

the dynamic stress intensity factor from experiments. Three-point-bend specimens of PMMA and AISI 4340 steel were loaded in a Dynatup 8100A drop-weight tower. The CGS fringe patterns resulting from the growing crack was captured using a high-speed camera (see Section 8.4 for details regarding the method). The crack speed was measured to be $0.25C_s$. Fringe patterns corresponding to a typical test are shown in the selected sequence of high-speed photographs in Fig. 8.14. Fringes observed in this technique are lines of constant gradient $\partial(\sigma_{11} + \sigma_{22})/\partial x_1$. The main advantage of the method over caustics is that it is a full-field technique; pointwise measurements over the complete field of view can be obtained; however, it suffers from the same drawback that due to the three dimensionality of the stress field, measurements must be made at $r/h > 0.5$. Introducing the dynamic stress field into Eq. 8.51 and rearranging the terms, it can be shown that if a K-dominant field is generated, then the ratio

$$Y_1^d(r, \theta) = \frac{m\lambda}{\Delta x_1} \frac{\sqrt{2\pi}}{hcF(v)} \frac{(r_d)^{3/2}}{\cos\dfrac{3\theta_d}{2}} \tag{9.2}$$

must be independent of r and be equal to the dynamic stress intensity factor. Krishnaswamy et al. (1992) evaluated this ratio along different radial lines from the crack tip and their results are shown in Fig. 9.8. The discrete symbols indicate Y_1^d calculated at a fringe order m and location (r, θ). Clearly, there is no region where the dynamic stress intensity factor dominates. The conclusion that the dynamic singular field does not dominate at $r/h > 0.5$ is not surprising in the wake of the analytical results

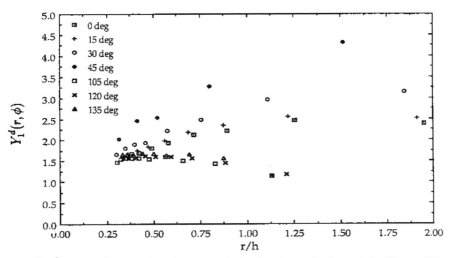

Figure 9.8 $Y_1^d(r, \theta)$ **vs** r/h **for a PMMA specimen, indicating that under dynamic loading conditions, CGS fringes can not be interpreted in terms of the singular stress field for** $r/h > 0.5$**. (Reproduced from Krishnaswamy et al. 1992.)**

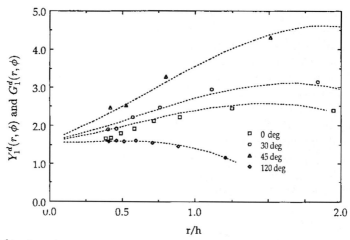

Figure 9.9 $Y_1^d(r, \theta)$ vs r/h **for a PMMA specimen, indicating that under dynamic loading conditions, CGS fringes can be interpreted in terms of the higher order transient stress field for** $r/h > 0.5$. **(Reproduced from Krishnaswamy et al., 1992.)**

of Ma and Freund (1986) and the experimental results of Ravi-Chandar and Knauss (1987) and Krishnaswamy and Rosakis (1990) obtained with the method of caustics.

However, the CGS fringes contain much more information than do the corresponding caustics. If the fringes are interpreted using the transient asymptotic field, a different result emerges. Interpreting the CGS fringes shown in Fig. 8.14 again, but this time with the transient crack tip stress field given in Appendix A, the equation for the bright fringes can be written as:

$$Y_1^d = \frac{m\lambda}{\Delta x_1} \frac{\sqrt{2\pi}}{hc_e F(v)} \frac{(r_d)^{3/2}}{\cos\dfrac{3\theta_d}{2}} = G_1^d(r, \theta; K_1^d, A_1, A_2 ... A_5) \qquad (9.3)$$

Krishnaswamy et al. extracted the parameters of the first six orders of the asymptotic expansion, $(K_1^d, A_1, ... A_5)$, by fitting the experimentally measured fringe data (left hand side of Eq. 9.3). The experimental results are shown in Fig. 9.9. Comparison of the radial variation of Y_1^d to the best theoretical estimate G_1^d is shown in this figure for various radial lines. The agreement is quite good, suggesting that interpretation of the CGS fringe patterns through a transient asymptotic expansion is appropriate. Since the analytical solution for the stress intensity factor is not known, comparison to theory is not possible. In an attempt to verify the efficacy of the data extraction procedure, Krishnaswamy et al. regenerated the fringe pattern from the estimated parameters. The comparison of the reconstructed fringes to the experimental fringes is shown in Fig. 9.10. Clearly, while the K-dominant analysis does not fit the experimental observation well, the higher order transient analysis appears to have captured the experimental data quite well, at least ahead

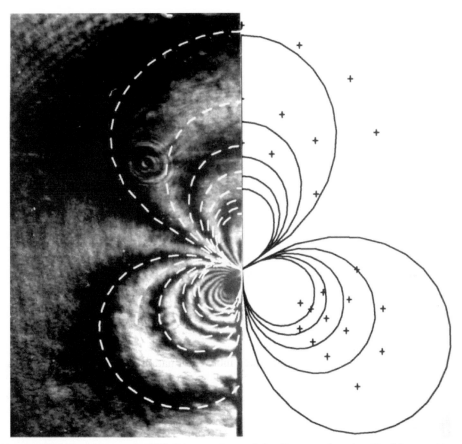

Figure 9.10 Comparison of CGS fringes observed in the experiment with fringe patterns reconstructed using the best estimates for the field parameters. (Reproduced from Krishnaswamy et al., 1992.)

of the crack. The conclusion from these experiments is that even though the dominance of the singular field is not established, by taking into account the development of the transient stress field over a large region near the crack tip, a consistent estimate of the field parameters can be obtained, and hence the dynamic stress intensity factor may be used as a fracture characterizing parameter.

Chapter 10

Dynamic Fracture Criteria

The energetic basis of fracture criterion to be used in dynamic problems was discussed in Chapter 5. It was also indicated that for practical applications, this fracture criterion is typically formulated in terms of separate criteria for initiation, growth and arrest of cracks. In this chapter, we describe experimental determination of these fracture criteria in different materials.

10.1 Criteria for Crack Initiation

Here attention is focused on experimental measurements of the dynamic crack initiation toughness in various materials. Recall that for a stationary crack under time-dependent loading, for small-scale yielding conditions, the crack tip stress and deformation field are determined by the dynamic stress intensity factor and that the initiation toughness is the critical value at initiation of crack growth, denoted by K_{Id}. The dynamic initiation toughness may depend on the loading rate and temperature. The loading rate appropriate to the crack tip region is characterized by $\dot{K}_I^{dyn}$ since all field parameters near the crack tip are proportional to $K_I^{dyn}(t)$. The main experimental task is then to characterize this material property, $K_{Id}(\dot{K}_I^{dyn}, T)$, the *dynamic crack initiation toughness*, under controlled loading and environmental conditions. We provide a survey of the many attempts made in characterizing dynamic crack initiation toughness.

Experiments aimed at characterizing $K_{Id}(\dot{K}_I^{dyn}, T)$ require a sharp initial crack generated by fatigue precracking or by arrest of a dynamically growing crack, a repeatable stress wave loading scheme with a well-characterized load history, and the ability to vary the rate of loading over a wide range. In addition, diagnostic schemes for monitoring the time variation of crack position and load are required, in order to determine the instant of crack initiation and the calculation of the dynamic stress intensity factor. While it was recognized quite early in the development of fracture mechanics that the onset of crack initiation would depend on the rate of loading, quantitative measurements had to wait for the development of experimental methods meeting the requirements listed above and analytical methods to identify the appropriate crack tip asymptotics for stationary and growing cracks under dynamic loads. The earliest attempt to determine the loading rate dependence of the crack initiation toughness was reported by Eftis and Kraft (1965); their results are reproduced in Fig. 10.1. They used single-edge-notched specimens of a carbon steel alloy, and determined the crack initiation toughness over a temperature range of -195 to $80°C$ and a loading rate in the range $1 \times 10 \le \dot{K}_I^{dyn} \le 1 \times 10^6$ MPa $\sqrt{m}/s$. They found that $K_{Id}(\dot{K}_I^{dyn}, T)$ was a decreasing function of $\dot{K}_I^{dyn}$ over this range. In order to extend the range of the data, they reinterpreted the experimental results on wide-plate tests performed by Videon et al. (1963) and obtained estimates of the possible rate dependence of initiation toughness over the higher rates of loading; the experimental measurements of Videon et al. were obtained for rapidly growing crack. Therefore, Eftis and Kraft (1965) displayed the estimated stress intensity factors as a function of the crack speed and not the loading rate. While the analysis based on elastostatics predated a complete understanding of the dynamic crack problem, and therefore, was not quantitatively correct, the trends described in their results—that the dynamic crack initiation toughness would first decrease slightly as the loading rate increased and then subsequently increase dramatically—are indeed intriguing and important. An accurate evaluation of the loading rate dependence of the dynamic crack initiation toughness and an understanding of the microstructural mechanisms responsible for such rate dependence are essential for the assessment of the safety of structures subjected to dynamic loading. In this section, we present a survey of investigations in dynamic crack initiation, its rate dependence and the fracture mechanisms responsible for rate dependence.

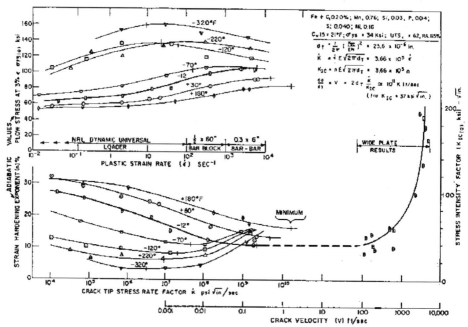

Figure 10.1 Effect of loading rate on the yield and fracture behavior of a high-strength steel. (Reproduced from Eftis and Kraft, 1965.)

10.1.1 Initiation of Cracks Under Short Duration Stress Pulses

Shockey and Curran (1973) evaluated crack initiation toughness at high strain rates in an ingenious experiment that utilized pulse loading. Thin disks of polycarbonate of 38 mm diameter and 3 mm thickness were impacted by a flyer plate arrangement (see Chapter 6 for a description of the experimental apparatus) at a speed of about 140 m/s; the short-duration compression loading pulse reflected as a tensile pulse from the specimen free surface and created numerous internal microcracks. The specimen was recovered and the dimensions of the microcracks were measured—this impact resulted in a number of penny-shaped interior cracks; they were able to measure 48 microcracks ranging in diameter from 14 microns to 5.5 mm. The same specimen was impacted once again at different speeds (14.3, 32.6 and 57 m/s) with a stress pulse of duration $\tau = 2.8$ μs and stress amplitude ($\sigma^* = 27.8, 60.3, 94.5$ MPa) in order that only some of these microcracks may grow under the second loading. They found that none of the cracks grew at the lower impact speeds; on the other hand, at the highest speed, all microcracks with an initial diameter greater than about 1 mm grew, but those with a diameter smaller than 0.71 mm did not grow. From a measurement of the tensile stress generated during the loading and the dimensions of the penny-shaped cracks that were initiated into growth, they estimated the crack initiation toughness, using a quasi-static analysis. Kalthoff and Shockey (1977) and Shockey et al. (1983a,b) re-examined this initiation problem by subjecting the results to a dynamic analysis. The analyzed configuration is represented in a meridional section in Fig. 10.2; the stress pulse is of magnitude σ^* and duration τ propagating with a speed C_d and interacting with a penny-shaped crack of radius a. The stress intensity factor for the axisymmetric problem of a penny-shaped crack under a quasi-static stress of magnitude σ^* was derived by Sneddon (1946): $K_I^{stat} = \sigma^* \sqrt{2a/\pi}$. The stress intensity factor for a penny-shaped crack under pulse loading has been evaluated by Chen and Sih (1977). Two different situations must be considered in the dynamic problem: 'short' cracks and 'long' cracks. The distance traveled by the dilatational wave in the duration τ of the loading pulse defines a characteristic length $\xi = C_d\tau$. If the characteristic length is very large, i.e. $2a \ll \xi$, we have a short crack or a long-pulse duration. In this case, the finite diameter of the penny-shaped crack influences deformation; the dynamic stress intensity factor quickly rises and overshoots the quasi-static value by about 25%, oscillates about and settles down at the quasi-static value as a result of repeated wave interactions with the finite crack boundaries. Thus, for the short cracks or long-pulse loading, the maximum stress intensity factor attained during the loading history is:

$$K_I^{dyn}\big|_{max} = 1.25 K_I^{stat} = 1.25\sigma^* \sqrt{\frac{2a}{\pi}} \tag{10.1}$$

On the other hand, if the characteristic length is very small, i.e. $2a \gg \xi$, we have a long crack or a short-pulse loading. During the time of loading pulse, the stress waves from any point on the crack do not reach the diametrically opposite points before termination of the loading pulse; the finite diameter of the penny-shaped crack is never felt completely by the crack and the situation corresponds more closely to an unbounded medium. The stress intensity factor increases with time as $t^{1/2}$ (see Eq. 4.19 for the plane-strain

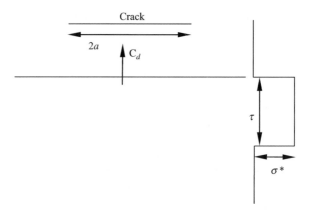

Figure 10.2 Interaction of a tensile stress pulse of magnitude σ^* and duration τ with a crack. The stress pulse travels with a speed C_d.

equivalent problem) and the peak stress intensity factor occurs not because of the interactions between the finite boundaries of the crack but because the loading pulse is of finite duration; Chen and Sih (1977) found the maximum stress intensity factor to be:

$$K_{\mathrm{I}}^{\mathrm{dyn}}\big|_{\max} = 0.59\sigma^* \sqrt{\pi\xi} \qquad (10.2)$$

Note that the maximum stress intensity factor is independent of the crack length and depends only on the duration of the loading pulse. It is now possible to determine the critical stress amplitude σ_c required for crack initiation by applying the crack initiation criterion: $K_{\mathrm{I}}^{\mathrm{dyn}}\big|_{\max} = K_{\mathrm{Id}}$. Then, from Eqs. 10.1 and 10.2, we get:

$$\sigma_c = \frac{K_{\mathrm{Id}}}{0.59\sqrt{\pi C_d \tau}} \quad \text{for long cracks or short pulses}$$
$$\sigma_c = \frac{K_{\mathrm{Id}}\sqrt{\pi}}{1.25\sqrt{2a}} \quad \text{for short cracks or long pulses} \qquad (10.3)$$

This analysis provided a rational interpretation of the experimental observations of Shockey and Curran (1973) and Kalthoff and Shockey (1977). Their results are shown in Fig. 10.3; in this figure, the stress amplitude, σ_c, is plotted on the ordinate and the crack length is plotted on the abscissa. Open symbols represent cracks that were not initiated and filled symbols indicate cracks that grow during the pulse loading. The lines drawn at the boundary between no growth and growth of cracks are based on the maximum stress intensity factor reaching the initiation toughness; the results also provide a good estimate of the dynamic initiation toughness of the polycarbonate material: $K_{\mathrm{Id}} = 2.2 \pm 0.2\,\mathrm{MPa\,m}^{1/2}$. This value is about 40% lower than the plane-strain fracture toughness determined from quasi-static experiments. Shockey et al. (1983a,b) estimated the rate of loading in this short-pulse projectile impact experiment to be about $\dot{K}_{\mathrm{I}}^{\mathrm{dyn}} = 10^7\,\mathrm{MPa\,m}^{1/2}/\mathrm{s}$. The short-pulse loading experiment, while difficult to implement because of the equipment requirements, provides an effective method for the determination of the dynamic crack initiation toughness. However, it is difficult to

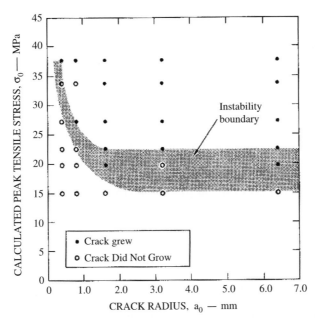

Figure 10.3 Data from Shockey et al. (1983a,b) indicating the variation of the critical stress amplitude required for crack initiation as a function of the crack radius. (Reproduced from Shockey et al., 1983a,b.)

vary the rate of loading over a large range and so its utility is primarily in the very high rate of loading conditions.

Shockey et al. (1983a,b) interpreted the loading rate dependence of the dynamic initiation toughness by arguing that while it was necessary for the stress intensity factor to reach a critical value, it was not sufficient; they postulated that the critical value of the stress intensity factor must be maintained for a minimum time for the fracture processes to develop completely. This idea can be justified only if specific kinetic processes that occur within the process zone on the same time scale as the applied loading are postulated. Such processes must depend on the fracture mechanisms that dictate crack growth and hence on the material, and even for the same material depend on whether a brittle or ductile fracture mechanism is triggered.

10.1.2 Loading Rate and Temperature Dependence of Crack Initiation Toughness

Costin et al. (1977) developed another experimental scheme to examine the loading rate dependence of dynamic crack initiation toughness. The experimental arrangement—an implementation of the Hopkinson bar apparatus—is shown in Fig. 6.7a. A tensile wave generated by detonation of an explosive charge traveled down the length of the specimen to reach the fatigue precrack and load it to failure within 25 μs. The load, $P(t)$, across the cracked specimen was measured with strain gages mounted on the specimen and the load point displacement, $\delta(t)$, was monitored with a moiré grid technique; the time variations of the load and crack opening displacement are shown in Fig. 6.8. Costin et al. (1977) then used the quasi-static analysis to evaluate the fracture toughness. Two different conditions were identified: first, when the condition of small-scale yielding was fulfilled, i.e. $R \geq 2.5(K_{\mathrm{I}}/\sigma_{\mathrm{Y}})^2$, the stress intensity factor is given by

$$K_{\mathrm{I}}^{\mathrm{stat}} = \frac{P}{\pi R^2} \sqrt{\pi R} f(R/D) \tag{10.4}$$

where R is the remaining uncracked ligament, D the diameter of the bar and $f(R/D)$ the geometric factor for the round-notched bar (Tada, 1973). Second, if the specimen size requirements for the plane-strain fracture toughness test were not met, an estimate of the J-integral for cracked round bars by Rice et al. (1973) was used to evaluate the stress intensity factor at crack initiation:

$$J = \frac{1 - \nu^2}{E} K_{\mathrm{I}}^2 = \frac{1}{2\pi R^2} \left[3 \int_0^{\delta_c} P(\delta) \mathrm{d}\delta - P\delta_c \right] \tag{10.5}$$

This quasi-static analysis was justified because the ligament dimensions in the cracked specimen were small; this claim was later shown to be acceptable by Nakamura et al. (1986) through a full-scale numerical simulation of the experiment. Costin et al. (1977) examined the initiation toughness and its dependence on temperature in a 1018 cold-rolled steel. Wilson et al. (1980) determined the temperature dependence of a 1020 hot-rolled steel. The loading rate in these experiments was fixed at a value of $\dot{K}_{\mathrm{I}}^{\mathrm{dyn}} > 2 \times 10^6$ MPa m$^{1/2}$/s. Their measurements of the temperature dependence of the dynamic crack initiation toughness are shown in Fig. 10.4. The fracture toughness obtained from quasi-static tests are also shown in this figure. There are a number of trends in these data that needs to be discussed further. First, the quasi-static experiments indicate brittle fracture by cleavage at temperatures below about $-150°$C with $K_{\mathrm{Id}} = 40$ MPa $\sqrt{\mathrm{m}}$ for both steels. Ductile fracture by void nucleation and growth was observed at temperatures above the transition temperature, with $K_{\mathrm{IC}} = 80$ MPa $\sqrt{\mathrm{m}}$ and $T_{\mathrm{NDT}} = -60°$C for the hot-rolled steel and $K_{\mathrm{IC}} = 110$ MPa $\sqrt{\mathrm{m}}$ and $T_{\mathrm{NDT}} = -110°$C for the cold-rolled steel. Second, the temperature dependence of the dynamic initiation toughness evaluated at a strain rate of about $\dot{K}_{\mathrm{I}}^{\mathrm{dyn}} > 2 \times 10^6$ MPa $\sqrt{\mathrm{m}}$/s was similar to the quasi-static toughness, but with the transition temperature for both alloys increasing to $T_{\mathrm{NDT}} = 30°$C. Third, the dynamic initiation toughness was dramatically lower than the quasi-static toughness below the transition temperature, but exceeded the low strain rate toughness at temperatures above 20°C. Finally, dynamic initiation toughness of both alloys was nearly the same over the entire temperature range tested.

From an examination of the fracture surfaces, Costin et al. and Wilson et al. identified that, associated with the transition temperature was a switch in the fracture mechanism from brittle cleavage to ductile fibrous fracture. Brittle fracture dictated by cleavage was induced at the carbide particles. Wilson et al. (1980) determined that the mean thickness of the grain boundary carbide plates was the same in both alloys and the mean spacing between carbide plates, favorably oriented to act as triggers for cleavage fracture, was also the same in both alloys. These microstructural features were used to justify the observed similarity in the dynamic initiation toughness between the two alloys in the brittle fracture

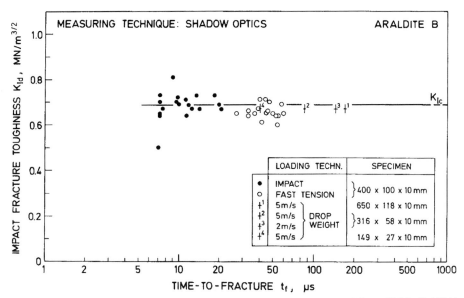

Figure 10.7 Dynamic crack initiation toughness of Araldite. (Reproduced from Kalthoff, 1986.)

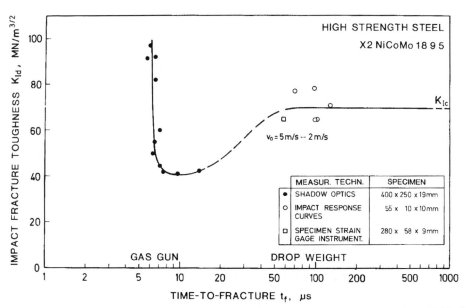

Figure 10.8 Dynamic crack initiation toughness of high-strength steel. (Reproduced from Kalthoff, 1986.)

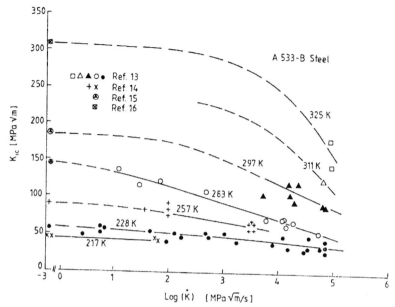

Figure 10.9 Dependence of the dynamic initiation toughness on the loading rate and temperature for A533B reactor grade steel. (Reproduced from Klepaczko, 1990.)

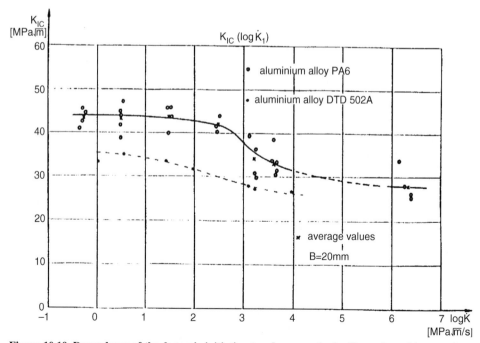

Figure 10.10 Dependence of the dynamic initiation toughness on the loading rate and temperature for two different aluminum alloys. (Reproduced from Klepaczko, 1990.)

the stress intensity factor, although the dynamic load measured from the Hopkinson bar apparatus was used. As seen in the figures, in all these tests, the initiation toughness was found to decrease with increasing loading rate and increase with increasing temperature, for the steel as well as the aluminum alloys.

In contrast, in another investigation, Rittel and Maigre (1995) used the compact compression specimen described in Chapter 6 (see Fig. 6.7c) to examine the strain rate dependence of the initiation toughness. The applied load and boundary displacements were measured with strain gages mounted on the incident and transmitter bars of the Hopkinson pressure bar apparatus. The strain gage signals were used to determine the time variation of the dynamic stress intensity factor through a path-independent integral described by Bui et al. (1992). The initiation and growth of the crack was monitored with two single wire fracture gages glued across the crack line, one on each side of the specimen. Rittel and Maigre (1995) evaluated the rate dependence of polymethylmetha-crylate (PMMA); the rate of loading in these experiments was determined to be in the range: $\dot{K}_I^{dyn}(t, v) = 1 \times 10^4 - 2 \times 10^5$ MPa $\sqrt{m}$/s, typical of the range used in the determination of initiation toughness in polymers. Their results indicated that the dynamic initiation toughness in PMMA increased from about 2 MPa $\sqrt{m}$ for quasi-static loading to low rates of loading $(\dot{K}_I^{dyn}(t, v) < 1 \times 10^4$ MPa $\sqrt{m}$/s), to about 13.5 MPa m$^{1/2}$ under dynamic loading at rates of about $\dot{K}_I^{dyn}(t, v) \sim 2 \times 10^5$ MPa $\sqrt{m}$/s; while these values are twice what is normally reported for this material (typically the quasi-static fracture toughness of PMMA is quoted to be around 1 MPa $\sqrt{m}$ although this depends significantly on the molecular weight), the trend of increasing toughness with loading rate corresponds well with observations in this and other polymers.

Owen et al. (1998) examined the rate dependence of crack initiation toughness in thin sheets (1.63, 2.03 and 2.54 mm) of 2024-T3 aluminum alloy. Single-edge-notched specimens (see Fig. 6.1) were loaded dynamically by a tension pulse in a split-Hopkinson tension apparatus. The strain gages in the incident and transmitted bars were used to evaluate the stress state in the specimen; assuming that the characteristic length ξ was very small, Owen et al. suggested that a quasi-static analysis was adequate for the determination of the dynamic stress intensity factor. They verified that this assumption was appropriate for small specimens by measuring the crack opening displacement directly and comparing it to the calculations based on the estimate of the dynamic stress intensity factor obtained from the strain gage measurements. The loading rates imposed in this experiment were in the range of $\dot{K}_I^{dyn}(t, v) = 1 \times 10^4 - 2 \times 10^6$ MPa $\sqrt{m}$/s. The results of Owen et al. are shown in Fig. 10.11. The plane-strain fracture toughness of this aluminum alloy was about 30 MPa $\sqrt{m}$. The aluminum alloy exhibited a monotonic increase from this quasi-static value at low loading rates, $\dot{K}_I^{dyn}(t, v) < 1 \times 10^4$ MPa $\sqrt{m}$/s, to a value of about 77 MPa $\sqrt{m}$ at the highest loading rate $\dot{K}_I^{dyn}(t, v) = 2 \times 10^6$ MPa $\sqrt{m}$/s. This is in contrast to the data shown in Fig. 10.10 from Klepaczko (1990); presumably both alloys are nearly the same and therefore exhibit a similar fracture response. Also shown in Fig. 10.11 is the rate of loading dependence of the crack initiation toughness in brittle polyester Homalite-100 (Ravi-Chandar and Knauss, 1984a); the plot is in normalized form of the data shown in Fig. 10.5 for comparison. The similarity between the rate dependence of these two vastly different materials is indeed striking. However, the underlying fracture mechanisms are significantly different in the metallic and polymeric materials;

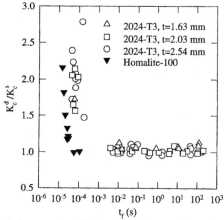

Figure 10.11 Variation of the crack initiation toughness with time-to-fracture for 2024-T3 aluminum alloy; comparison to the behavior of Homalite-100 is also shown. The time to fracture is inversely related to the loading rate. (Reproduced from Owen et al., 1998.)

hence the similarity must arise from some other common attribute. It appears that the inertial effect in the build-up of stress fields near the crack could be the main contributor to the loading rate dependence of the initiation toughness as discussed in the model by Liu et al. (1998).

10.2 Dynamic Crack Arrest Criterion

The general formulation of the dynamic crack arrest criterion has been described in Chapter 5. In this section, a discussion of the experimental investigations aimed at characterizing the crack arrest behavior is presented. Recall that dynamic crack arrest toughness, $K_{\mathrm{Ia}}(T)$, was defined as the smallest value of the dynamic stress intensity factor for which a growing crack cannot be maintained. As described earlier, a conservative design can be implemented in practice based on crack arrest criteria alone. If the dynamic stress intensity factor of a crack in any structure never exceeds $K_{\mathrm{Ia}}(T)$, a growing crack can never be sustained and hence the design is not susceptible to dynamic crack growth! So, the main focus of experiments has been the determination of the crack arrest toughness. In this section, we describe first the studies aimed at identifying the crack arrest criterion, and then follow this with a description of the standard test method for the determination of the arrest toughness. Finally, we describe strategies adopted for the design of fracture critical structures.

10.2.1 Development of the Crack Arrest Criterion

Early attempts at characterizing crack arrest toughness as a material property were actually motivated by difficulties in determining the dynamic initiation toughness

(Crosley and Ripling, 1969). In their experiments aimed at determining the loading rate dependence of the dynamic crack initiation toughness of A533B steel, Crosley and Ripling found that the state of crack arrest was more easily reproduced than the state of crack initiation. This is readily understood by considering that while crack initiation toughness is influenced significantly by the bluntness of the initial crack, crack arrest always proceeds by the deceleration of a natural crack and is independent of how the crack was produced. Hence from an experimental point of view, it is easy to create appropriate conditions for crack arrest than for initiation. However, the crack arrest experiment is inherently a dynamic experiment and time-resolved measurements of the crack position and load are generally required in order to determine the crack arrest toughness. As was the case of the crack initiation problem, early investigations on crack arrest relied upon a static analysis for the interpretation of experimental measurements in terms of the stress intensity factor at arrest. However, with the development of the various diagnostic techniques described in Chapters 7 and 8, many careful studies have examined the determination of the crack arrest criterion. Many different types of specimen and loading have been used to generate arrest of a rapidly growing crack; these have been interpreted based on static or dynamic analysis. ASTM Publications STP 627 (1977) and 711 (1980) contain detailed discussion of the development of the many techniques. Here we provide a discussion of two experiments that provide an overview of the determination of crack arrest toughness.

The key ingredient necessary in crack arrest experiments is that at some time during the test history the crack arrest condition is satisfied

$$K_{\mathrm{I}}^{\mathrm{dyn}}(t) \leq K_{\mathrm{Ia}}(T) \text{ for } t > t_{\mathrm{a}} \tag{10.6}$$

where t_{a} is the time of crack arrest. This can be accomplished in one of the three ways: first, by performing the test in a configuration where $K_{\mathrm{I}}^{\mathrm{dyn}}(t)$ is a decreasing function of time; a number of different configurations have been used to accomplish this. From Eq. 5.15, it is seen that if a wedge-loaded specimen is used, as the crack extends, the stress intensity factor will decrease and eventually satisfy Eq. 10.6; this is the basis of the ASTM Standard test. Crosley and Ripling (1980), Hoagland et al. (1977), Kalthoff et al. (1977) and others also used this approach. Kobayashi et al. (1977) used a single-edge-notched specimen but with loading that decreased linearly away from the crack tip resulting in a decreasing $K_{\mathrm{I}}^{\mathrm{dyn}}(t)$ with crack extension. Ravi-Chandar and Knauss (1984b) studied crack arrest in the electromagnetic loading scheme by applying a short-duration pulse loading. The second method of generating crack arrest is by increasing the crack arrest toughness along the crack path; this is the basis of the Robertson test (1953) where the cracked end of the specimen is held at a very low temperature while the other end is kept at a significantly higher temperature. This idea has also been used in the wide-plate tests at the National Institute of Standards and Technology (Pugh et al., 1988). As the crack extends, it encounters a material with a higher crack arrest toughness and once again inequality in Eq. 10.6 is satisfied at some point along the crack path. Lastly, duplex crack arrest specimens have also been used: here a fast crack is started in a material with a low toughness that then grows into the second material of higher crack arrest toughness and hence arrests. Hoagland et al. (1977), Dally and Kobayashi (1978) and others have used this approach to crack arrest studies. All these investigations have been instrumental in identifying that the dynamic crack arrest toughness can be established as a material property, dependent on temperature, and in establishing that a dynamic analysis is necessary to interpret the experimental measurements reliably in terms of the arrest toughness.

Kalthoff et al. (1977) performed an illuminating series of experiments to determine the crack arrest toughness as well as the necessity of performing a dynamic evaluation. Specimens of Araldite B, a brittle epoxy, were used in a rectangular double-cantilever beam configuration. The loading was generated by wedging the crack as indicated in Fig. 10.12. The initial crack was made blunt, with varying radius. The wedge opening displacement at the instant of crack initiation was determined through a clip gage; since dynamic effects are not involved prior to crack initiation the stress intensity factor at the onset of crack growth, labeled K_{Iq} to distinguish it from the initiation toughness, can be determined by static analysis. While the dynamic crack initiation toughness for Araldite B is about 0.79 MPa $\sqrt{m}$ (see Fig. 10.7), by using blunt notches of different notch radii, Kalthoff et al. (1977) were able to vary K_{Iq} in the range 0.7–2.5 MPa $\sqrt{m}$. For the propagating crack, Kalthoff et al. recorded caustic patterns from the crack tip on a Cranz-Schardin type camera. The high-speed camera was triggered by interruption of a laser beam by the crack as it grew from the initial blunt notch; thus the initial phase of crack growth was not captured and the early time history of the dynamic stress intensity factor was not determined. The variations of the dynamic stress intensity factor, K_I^{dyn}, and the crack speed, v, obtained from the caustic measurements with crack position are shown in Fig. 10.13. Also indicated in this figure is the estimate of the stress intensity factor, K_I^{stat}, calculated from the measured crack position and crack opening displacement. In particular, the static calculation of the stress intensity factor with arrested crack length and measured crack opening displacement yields the static estimate for the crack arrest toughness: K_{Ia}^{stat}. A number of important features in the experimental results must be noted. First, K_I^{dyn} exhibits a rapid drop not captured in the data recorded, but settles down at a constant value for much of the crack growth period and then gradually drops down to the arrest value. Second, the crack speed is nearly constant during the period that K_I^{dyn} is constant; deceleration of the crack appears as K_I^{dyn} decreases, but always lags the latter. The reasons for this are not completely understood. Third, for all the experiments shown, the crack comes to a stop when K_I^{dyn} reaches a constant value of about 0.7 MPa m$^{1/2}$; thus, this value must be considered to be the crack arrest toughness, K_{Ia}. Beyond crack arrest, K_I^{dyn} continues to drop; the time variation of K_I^{dyn} for the arrested crack is shown in

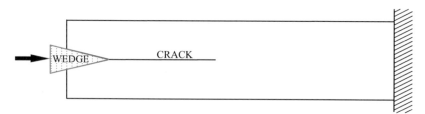

Figure 10.12 Wedge-loaded rectangular double-cantilever beam configuration.

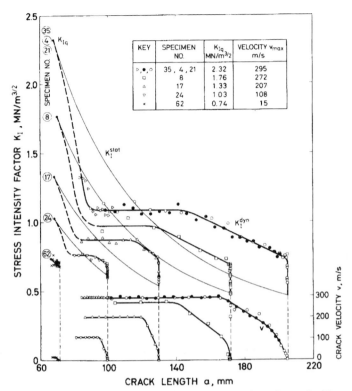

Figure 10.13 **Variation of the dynamic stress intensity factor and crack speed with crack position in a wedge-loaded double-cantilever beam specimen. (Reproduced from Kalthoff et al., 1977.)**

Fig. 10.14. In this figure, K_I^{dyn} normalized K_{Ia}^{stat} is plotted as a function of time. Clearly, K_I^{dyn} for the arrested crack oscillated about K_{Ia}^{stat} and eventually settled down at this value; however, crack arrest did not occur at K_{Ia}^{stat}, but at K_{Ia} as indicated in Fig. 10.13. K_{Ia}^{stat} is always lower than K_{Ia} in all the experiments and furthermore, K_{Ia}^{stat} depends on the initial notch bluntness! Finally, from the comparison shown in Fig. 10.13 between K_I^{dyn} and K_I^{stat}, the static analysis is never appropriate in this test, except at very long times when the crack has been arrested and all the oscillations have died out.

In another series of experiments, Ravi-Chandar and Knauss (1984a) used the electromagnetic loading method to examine crack initiation and arrest in the same test. In this configuration of the pressurized semi-infinite crack geometry the crack was initiated, arrested and reinitiated, all *before the arrival of waves reflected* from the far boundaries of the specimen; thus, in this experiment there are no geometrical influences on crack arrest. Crack arrest was achieved by loading the crack faces dynamically with a load just sufficient to start crack growth. The shape of the loading pulse was designed to be long enough to generate crack growth, yet so short that after about a few millimeters of crack growth, the crack tip was unloaded. A double-trapezoidal pulse with each pulse lasting

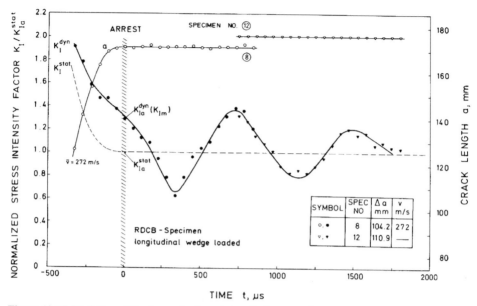

Figure 10.14 Variation of the dynamic stress intensity factor and crack speed with time after crack arrest. (Reproduced from Kalthoff et al., 1977.)

about 70 μs was used in the loading. The results from this series of tests on Homalite-100 are shown in Fig. 10.15; the variation of K_I^{dyn} and the crack position with time are shown in the figure. The starter crack was made using a razor blade as a wedge, which led to variations in the initiation stress intensity factor. From Fig. 10.15 it is evident that when K_I^{dyn} reached the dynamic initiation toughness K_{Id}, the crack began to grow at a constant speed; associated with the crack extension is a drop in the stress intensity factor (as dictated by Eq. 4.56). As soon as the arrest condition in Eq. 10.6 is satisfied, the crack is arrested. Note that, in all the experimental results shown in Fig. 10.15, at around 70 μs the stress intensity factor is indeed always the same within the accuracy of the measurement; the crack arrest appears quite consistently at $K_I^{dyn} = 0.4$ MPa m$^{1/2}$. This is about 11% lower than K_{IC} for this material (see Fig. 10.5). Since these tests were conducted in an infinite specimen geometry, there are no reflected stress waves and hence no oscillations in the stress intensity factor. Clearly, the crack arrest toughness can be taken to be $K_{Ia} = 0.4$ MPa m$^{1/2}$. Continued loading from the second pulse of the loading reinitiated the arrested crack at K_{Id}.

From these careful sets of experiments described above, it is easy to conclude that K_{Ia} is an appropriate measure of the crack arrest toughness and that a dynamic analysis is essential in order to determine this parameter from experiments. However, for standard tests, it is essential to have a simple experimental procedure that utilizes conventional laboratory apparatus. Kalthoff et al. (1977) showed that in some specimen configurations such as the compact tension specimen, the dynamic effects may be small and a quasi-static analysis may be used conservatively; this geometrical configuration has been implemented in the ASTM standard test procedure described in Section 10.2.2.

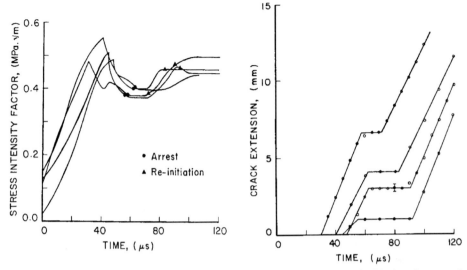

Figure 10.15 Variation of the dynamic stress intensity factor and crack speed with time from crack initiation and arrest experiments. (Reproduced from Ravi-Chandar and Knauss, 1984a.)

10.2.2 ASTM Standard Method for Crack Arrest

Based on round-robin tests and an accumulation of data, a standard test procedure, the ASTM E-1221 Standard, has been established that describes the determination of crack arrest toughness in ferritic steels. A brief summary of the test method is provided here. The schematic diagram of the test arrangement under this standard is shown in Fig. 10.16. In this arrangement, the geometry of a compact tension specimen is used; however, loading is generated by introducing a split pin into the hole in the specimen and forcing the pins apart

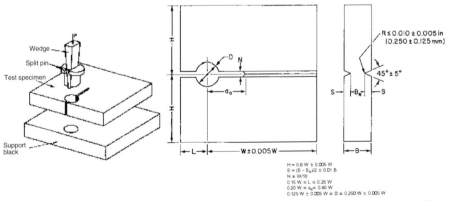

Figure 10.16 Specimen and loading configuration for the determination of crack arrest toughness. (Reproduced from ASTM E-1221 Standard.)

by a wedge with a force P; this configuration is called the compact crack arrest (CCA) specimen. Side-grooves are introduced to guide the crack along the plane of symmetry. A rapid crack growth arrest sequence is generated by forcing a wedge; as the crack grows away from the hole, the stress intensity factor drops quickly (see the discussion in Section 5.6.) and hence results in arrest of the crack. The preparation of the starting notch is quite important to the successful performance of this standard test. In low to intermediate strength alloys, it is suggested that the crack line be embrittled by depositing a weld along the crack line; other forms of the starter notch include a quench-embrittled Chevron notch, and a fatigue-precracked and overloaded crack. The starter notch dictates the level of the stress intensity factor at which crack initiation occurs; the extent of subsequent crack growth segment is dictated by this value. If the stress intensity factor at initiation is low enough, arrest of the crack can occur at an appropriate length within the compact specimen. The dynamic run arrest sequence that this specimen experiences under the wedge load clearly indicates the need for a dynamic analysis of the problem. However, according to the ASTM standard, a static analysis is considered to be appropriate; thus from a recording of the maximum load, P, and the crack mouth opening displacement, δ, the stress intensity factor can be determined. The load and crack mouth opening displacement at the onset of crack initiation are labeled P_0 and δ_0, while the corresponding values soon after crack arrest are labeled P_a and δ_a; typically the arrest values oscillate with time, but the ASTM standard assumes that the values measured at 2 ms after crack arrest do not differ significantly from the values measured at 100 ms after crack arrest. The stress intensity factors corresponding to crack initiation K_0 and arrest K_a are then calculated from a static analysis through the following equation

$$K_I = E\delta f\left(\frac{a}{W}\right)\sqrt{\frac{B}{B_N W}} \qquad\qquad (10.7)$$

where δ is the appropriate crack mouth opening displacement and E the modulus of elasticity; other geometrical quantities are defined in Fig. 10.16. The geometric correction factor $f(\alpha)$ is given by:

$$f(\alpha) = (1 - \alpha)^{0.5}(0.748 - 2.179\alpha + 3.56\alpha^2 - 2.55\alpha^3 + 0.62\alpha^4) \qquad\qquad (10.8)$$

The stress intensity factor at arrest is taken to be the crack arrest toughness, K_{Ia}, if the following conditions are met: (i) the crack growth during the run arrest segment must be at least greater than the plane-stress plastic zone corresponding to K_0 and greater than twice the side-groove slot width; (ii) the thickness $B \geq (K_a/\sigma_{yd})^2$, and (iii) the unbroken ligament must be larger than 0.15W and $1.25(K_a/\sigma_{yd})^2$. The dynamic yield strength, σ_{yd}, is taken to be 205 MPa (30 ksi) larger than the static yield strength of the material. The standard test procedure also restricts the maximum crack mouth opening that can be attained in one continuous loading; if this value is reached, the specimen must be unloaded and reloaded with the maximum crack opening displacement increased to a specified level at each reloading. Further details can be found in the ASTM E-1221 Standard.

10.2.3 Application of the Crack Arrest Criterion

Once the crack arrest criterion has been postulated as in Eq. 10.6, and the crack arrest toughness determined as an appropriate material property in the range of temperatures of interest, application of the criterion is quite straightforward. For example, according to Milne et al. (1988) the R6 procedures for failure assessment could be used to evaluate flaw criticality under dynamic loading as long as the material properties are evaluated at the appropriate rates and quasi-static analysis of the stress intensity factors is appropriate; however, if inertial effects become important, such procedures become invalid. But, if the dynamic stress intensity factor variation is calculated analytically or through numerical simulations, evaluation of crack arrest is simply through the use of Eq. 10.6.

Two different design approaches have been used in industrial practice. The first one is based on the selection of a material with a large enough crack arrest toughness K_{Ia} that the structure should never experience a stress intensity factor (dynamically) that exceeds this value; hence crack initiation and growth do not occur in the structure. This is similar to the minimum specified fracture toughness in the ASME Boiler and Pressure Vessel Code, however, a requirement based on crack arrest toughness will be even more conservative since $K_{Ia} < K_{Id}$. While this stringent requirement is suitable for structures that must be designed to be damage-resistant, high performance structures and aging structures that have already accumulated flaws in service must be evaluated with a damage-tolerant approach. Therefore, the second approach to the critical design problem has been to perform a dynamic analysis of any particular design; in this approach crack initiation and growth of pre-existing cracks are allowed, but the extent of their growth is controlled through judicious design of the structure and by the placement of crack arresters, deflectors or tear straps; this has been the practice in ship and airplane industries. The design of the stiffeners or arrestor plates, however, requires a detailed fracture mechanics analysis of the dynamic loading and the resulting crack growth. While dynamic analysis of the crack arrest strategies such as these have been shown for a long time (see for example, Wade and Kobayashi, 1970), in most industrial practice, this is currently accomplished through scale model and full-scale tests. Two examples are provided in the following discussion in Figs. 10.17 and 10.18.

In Fig. 10.17, the use of an arrester plate is indicated, typical of applications in ship building and more recently in pipelines. For a hull made of material A with a crack arrest toughness K_{Ia}^A, a strip of material B with a crack arrest toughness $K_{Ia}^B \gg K_{Ia}^A$ is inserted by welding in a patch as indicated in Fig. 10.17; material B may simply be an alloy of slightly different composition than material A. Alternatively, a patch of material B or even the same material may be superposed on material A, much as a stringer attached to an aircraft skin; the stiffener can be welded or fastened with rivets, but essentially plays the same role as the welded plate in Fig. 10.17. The concept behind this method of crack arrest is the following: as the crack extends, let the stress intensity factor increase monotonically as indicated in the inset graph in Fig. 10.17; note that it is the dynamic stress intensity factor that must be calculated, either in some closed form approximation or through a numerical simulation of the problem. As the crack extends into material B, it suddenly encounters a material with a much higher crack arrest toughness; since $K_I^{dyn} < K_{Ia}^B$, the condition in

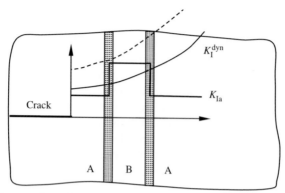

Figure 10.17 Crack arrest through an arrestor strip. A crack running from zone A into the arrestor zone encounters a material with higher arrest toughness in zone B and therefore arrests. In practice, the strap of material B is welded to material A.

Eq. 10.6 is satisfied and the crack arrests within material B. It must be noted that a dynamic analysis is necessary, for if a subsequent loading is applied on the same crack, perhaps from stress waves reflected from other parts of the body, it is possible to reinitiate the crack if these waves provide a stress intensity factor history indicated by the dotted line in Fig. 10.17.

The second example we describe is from applications in aircraft design. A typical aircraft fuselage is made of a thin metallic skin and reinforced by riveted longitudinal stringers and circumferential frames. The skin itself contains a riveted longitudinal lap splice joint. Longitudinal cracks appear in such structures from fatigue loading; small cracks grow at rivet sites in close proximity to each other and as they approach each other this *multi-site damage* can result in triggering catastrophic axial crack propagation along the lap splice joint which can then grow in an uncontrolled manner; the circumferential stiffeners do carry some of the load as well as provide additional resistance to crack growth as discussed in the previous paragraph, but the stiffeners eventually break. An effective method of arresting these axial cracks is through the use of tear straps as shown schematically in Fig. 10.18. Tear straps are of the same thickness as the skin, but are narrow strips that are attached to the skin at every circumferential frame and midway between the frames. The action of the tear strap can be considered in the following manner: analysis of the problem of an axial crack under internal pressure indicates that as the crack extends, the stress intensity factor must increase; however, as the crack approaches the tear strap/frame, since these parts share some of the load, the crack tip gets unloaded as shown in the inset in Fig. 10.18. At the same time, due to the increased cross-section to be cracked at the tear strap, the crack arrest toughness increases sharply. Thus, the crack cannot penetrate through the tear strap; however, due to pressure loading, a bulge develops on the skin, which results in a mixed-mode loading on the arrested crack. Under suitable conditions, the crack turns to the circumferential direction. This leads to an opening of a flap on the fuselage and hence depressurization of the cabin and a decrease in the driving force for crack extension. In order to implement this concept, the practice has

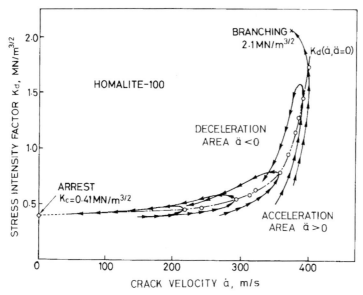

Figure 10.21 Crack growth toughness in Homalite-100. The toughness was shown to depend on whether the cracks were accelerating or decelerating. Contrary to the observations of Dally (1979) decelerating cracks experience a higher stress intensity factor. (Reproduced from Arakawa and Takahashi, 1991.)

there is a limiting speed at 335 m/s; at this speed, even though the stress intensity factor increases by more than a factor of 2, the crack speed does not change within the precision of the measurement; this is contrary to the one-to-one relationship exhibited in Eq. 5.8. Arakawa and Takahashi (1991) also found that the $K-v$ relationship depended on whether the crack was accelerating or decelerating, but their trend was exactly the opposite of that found by Dally. The results of Arakawa and Takahashi (1991) are shown in Fig. 10.21.

Kobayashi and co-workers performed numerous crack growth experiments on specimens of Homalite-100; they used SEN, DCB, three-point bending and other configurations with both quasi-static and dynamic (drop-weight tower) loading conditions. They also used dynamic photoelasticity and a Cranz–Schardin multi-spark camera to determine the crack growth criterion. Their results for the $K-v$ relationship are shown in Fig. 10.22. The scatter in the experimental data was extremely large and attributed to variations in the material and influence of specimen geometry and errors in crack speed measurement; it is also possible that a K-dominant field was not established in the region where the isochromatic fringe data were analyzed. Mall and Kobayashi (1978) concluded that it was difficult to establish a unique $K-v$ relationship through interpretation of their results.

Kalthoff (1983) used the method of caustics to determine the dynamic crack growth criterion in Araldite B. His earlier experiments indicated large scatter in the $K-v$ relationship. In order to sort this out, he performed three sets of experiments on the same

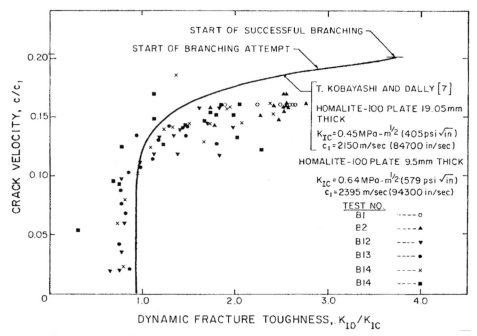

Figure 10.22 Crack growth toughness in Homalite-100. Large scatter in the data are observed in specimens of different geometry. (Reproduced from Mall and Kobayashi, 1978.)

batch of the specimen material in order to remove any concerns regarding material variability. In each set of experiments, different specimen geometry was used: one set of experiments was performed on a rectangular DCB specimen; the second set was on a single-edge-notched specimen. In the third set the specimen had two segments: a starting segment that had the same dimensions as the DCB and a second segment that had the dimensions of the SEN specimen. The $K-v$ relationship corresponding to these experiments are shown in Fig. 10.23; the specimen geometry corresponding to all three kinds of specimens are also shown in the inset. The upper curve in Fig. 10.23 corresponds to the DCB specimen and the lower curve corresponds to the SEN specimen. Kalthoff found that for the DCB/SEN specimens of the third set, the data from the DCB segment fell on the upper curve and that from the SEN segment fell on the lower curve. Thus Kalthoff demonstrated that there was a clear specimen geometry dependence on the $K-v$ relationship.

Ravi-Chandar and Knauss (1984c) used the electromagnetic loading device to investigate the dynamic crack growth criterion in Homalite-100. Since the loading corresponds to a pressure-loaded semi-infinite crack in an unbounded medium, this loading scheme should provide the dynamic crack growth criterion without the complications of specimen geometry dependence. They used a rotating mirror high-speed camera and captured the crack tip caustics right from crack initiation. In a startling departure from the previous measurements, they found that while the dynamic stress

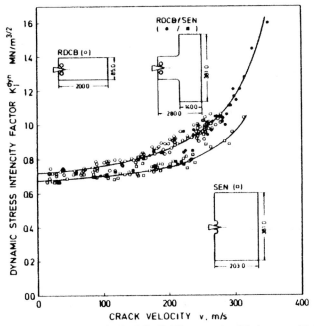

Figure 10.23 Crack growth toughness in Araldite B. The two sets of data were obtained on the same material, but in different loading geometries. Upper curve corresponds to the rectangular double-cantilever beam or the rectangular double-cantilever beam with an enlarged end and the lower curve corresponds to a single-edge-notched specimen. (Reproduced from Kalthoff, 1983.)

intensity factor continued to vary subsequent to crack initiation, the crack speed remained constant within experimental accuracy. Their results are reproduced in Fig. 10.24. The open symbols in the figure identify the stress intensity factor at crack initiation and the horizontal lines indicate the range of $K_{\mathrm{I}}^{\mathrm{dyn}}$ attained by the crack running at the same speed. At the highest load levels, they observed crack branching as indicated in Fig. 8.7. Also shown in Fig. 10.24 is the trend data from Dally (1979) for Homalite-100. From these experimental observations on the dynamic crack growth criterion, a few things are apparent. First, the driving force required to propagate a dynamic crack rapidly increases as the crack speed increases; however, this relationship may not be unique. Dependence of the $K-v$ relationship on specimen geometry, crack tip acceleration and loading as well as initial conditions have been shown through many experiments. Second, a limiting crack speed that is roughly about half the Rayleigh surface wave speed has been observed in almost all experimental investigations. Third, as the driving force is increased, cracks typically split into two or more branches, with each propagating at a high speed along its own path. Ravi-Chandar and Knauss (1984b,d) provided a mechanistic explanation for the nonuniqueness in the $K-v$ relationship, the appearance of the limiting crack speed and the appearance of crack branching based on a microcracking model of the fracture process; physical aspects of fracture process are discussed in Chapter 11.

Notwithstanding the nonuniqueness addressed above, for practical purposes, one can introduce an 'averaged' $K-v$ relationship, such as the one suggested by the dotted line in

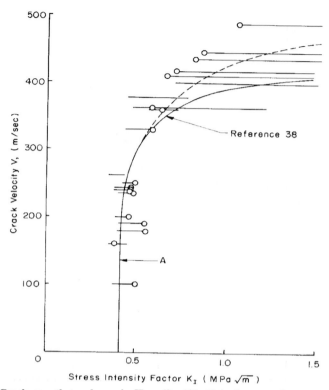

Figure 10.24 Crack growth toughness in Homalite-100. Note the nonunique response observed: while the stress intensity factor changes, the crack speed remains unaltered. The solid line (identified as Reference 38) is from Dally (1979). (Reproduced from Ravi-Chandar and Knauss, 1984c.)

Fig. 10.24 in order to evaluate the extent of crack growth in any particular experiment. While such a representation is a gross approximation, it is useful in obtaining an estimate of the extent of crack growth that might occur for use in conservative design practice. However, further work is needed in this area to determine the appropriate dynamic crack growth criterion for brittle materials. Some recent work on generating atomistic and lattice models as well as cohesive zone models of the process zone effects are described in Chapter 12.

10.3.2 Crack Growth Toughness in Ductile Materials

There have been fewer investigations of the dynamic crack growth toughness in ductile materials, possibly due to the difficulties associated with diagnostic methods that must be used on these materials. Kobayashi and Dally (1980) investigated the $K-v$ relationship in 4340 steel specimens; the specimens were heat treated for 1 h at 900°C, followed by an oil quench and tempering at 370°C. Wedge-loaded compact tension specimens similar in design to the CCA specimen were used. The crack tip state was determined with a

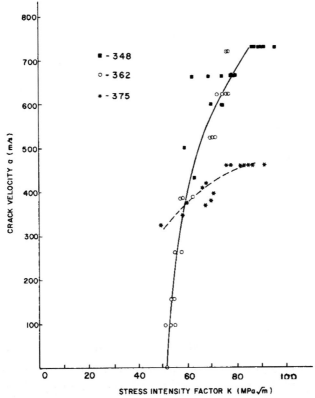

Figure 10.25 Crack growth toughness in a 4340 steel. (Reproduced from Kobayashi and Dally, 1980.)

photoelastic coating placed on the steel specimen. The variation of the dynamic crack growth toughness with crack speed determined from three experiments is shown in Fig. 10.25. The general trends in the K–v relationship are somewhat similar to that observed in the nominally brittle materials. In the early stages, a very small change in K_I^{dyn} results in a large change in the crack speed. Apparently, the specimen marked 375 in this figure was heat treated differently and hence exhibited a significantly lower initiation threshold and a lower limiting crack speed. Dahlberg et al. (1980) and Brickstad (1983) measured crack position with a high-speed camera or potential drop method, but calculated K_I^{dyn} through finite element analysis incorporating boundary measurements of loads. Dahlberg et al. (1980) examined 0.5 mm thick sheets cold-rolled and hardened carbon steel and observed significantly larger experimental scatter than Kobayashi and Dally (1980). They concluded that the large scatter is due to large errors in crack speed measurements and that if the measurements were performed in experiments where K_I^{dyn} did not vary much, the scatter would be reduced significantly. Brickstad (1983) made

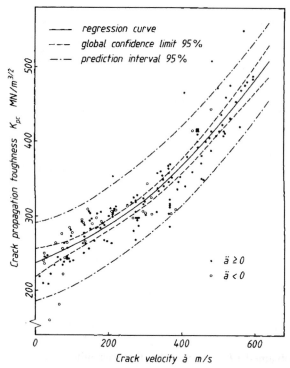

Figure 10.26 Crack growth toughness in a high-strength carbon steel. No obvious dependence on acceleration was noted. (Reproduced from Brickstad, 1983.)

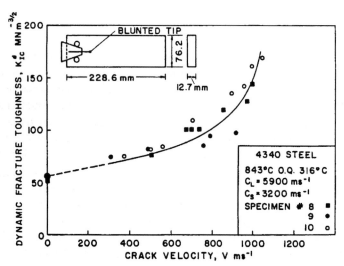

Figure 10.27 Crack growth toughness in 4340 steel. (Reproduced from Rosakis et al., 1984.)

measurements on a high-strength carbon steel. Single-edge-notched specimens were used; the crack position was monitored with a potential drop measurement (see Section 6.3). The boundary forces and displacements were measured and used as input to a finite element simulation in order to calculate K_I^{dyn}. Brickstad's estimate of the $K-v$ relationship is shown in Fig. 10.26. The general trend is similar to that shown by Kobayashi and Dally (1980). A large scatter in the data is observed; however, unlike the case of nominally brittle materials, results for accelerating and decelerating cracks are indistinguishable. Rosakis et al. (1984) evaluated the dynamic crack growth toughness of 4340 steel, heat treated to 843°C, followed by an oil quench and tempering at 316°C for 1 h. They used a DCB specimen and the method of caustics to measure the instantaneous K_I^{dyn} and crack position. Their results are shown in Fig. 10.27. A monotonic increase in K_I^{dyn} with crack speed v was observed. Thus, unlike the case of brittle materials, experimental characterization of the dynamic crack growth toughness, through both direct measurements and combinations of measurements and analysis have resulted in suggesting that a unique $K-v$ relationship is indeed possible in metallic materials.

The contrast in response between nominally brittle and ductile materials in the $K-v$ relationship must be attributed to the deformation and failure mechanisms that occur in the fracture process zone. In ductile materials, dissipation associated with plasticity is a significant fraction of the overall energy expended by the propagating crack and must be taken into account in dynamic fracture modeling. Lam and Freund (1985) performed a numerical simulation of a dynamically growing crack in a non hardening material, imposing a critical crack tip opening angle as the criterion for crack growth. From this simulation, they obtained a $K-v$ relationship that was quite similar to the results exhibited in Fig. 11.27; clearly, the rapid increase in dynamic fracture toughness with crack speed is attributed to the work of plastic deformation. For brittle materials, on the other hand, one must look into the particular details of the fracture process. These are addressed in the next chapter.

Chapter 11

Physical Aspects of Dynamic Fracture

In the previous chapters, analysis and experiments of dynamically loaded and propagating cracks have been discussed. Formulations of unified as well as ad hoc failure criteria have also been described. While the theory has been able to provide the time variation of the dynamic stress intensity factor as well as other field quantities, the ad hoc failure criteria discussed in Chapter 10 are not fully quite adequate in capturing the dynamic response of nominally brittle materials. The reason for this is quite simple: continuum fracture mechanics theory has focused on analyzing the 'outer problem' of determining the stress, deformation and energy flow in a region near the crack tip; this has been a fruitful exercise in that engineering structures at the large scale can be designed to be fracture-critical. However, in order to determine the fundamental failure character-istics, one needs to examine the 'inner problem' of understanding, characterizing and modeling the failure processes that actually lead to energy dissipation. The fact that the fracture processes have dynamics of their own—as seems to be the case with nominally brittle materials—leads to an interesting array of response of these materials. Careful experimental observations form the basis for developing an understanding of the fundamental physics and mechanics of the inner problem in dynamic fracture. From the pioneering experiments of Hopkinson (1901) and Schardin and Struth (1938), to the spurt of dynamic fracture activities of the 1970s and 1980s, there is a wealth of experimental observations of dynamic fracture phenomena at various scales. *There are a number of crucial experimental observations* and theoretical models and numerical simulations of the fracture process have to be constructed and performed with due attention to these observations.

In this chapter, we shall be concerned primarily with the materials that are characterized as nominally brittle—inorganic glasses, ceramics, and organic polymers in their glassy state would be included in this class of materials; the elastodynamic theory works well for ductile materials that possess a unique K_I-v relation and do not exhibit crack branching. Early dynamic fracture experiments were performed mostly in inorganic glasses (most of this work by Schardin and co-workers is summarized by Schardin (1959) and Kerhkof (1973)), but much of the recent experiments exploring dynamic fracture have been conducted in organic polymers (Kobayashi and Mall, 1978; Dally, 1979; Kalthoff, 1983; Ravi-Chandar and Knauss, 1984a–d). There have been very few dynamic fracture investigations in brittle crystalline materials (Bowden et al., 1967; Field, 1971; Cramer et al., 1997), probably due to difficulties associated with conducting the experiments. The concentration of effort on brittle organic glasses has been driven mainly by the possibility which these materials offer to measure crack tip stress or deformation fields through the standard techniques of experimental mechanics. Results of these experiments pertaining to the physics of dynamic fracture are summarized here to provide a mechanistic basis for understanding the fracture process.

11.1 Limiting Crack Speed

The energy balance described in Eq. 5.8 can be rewritten by combining all of the velocity dependence into one term so that:

$$g(v/C_R)K_I^0(t, l(t), 0) \approx \left(1 - \frac{v}{C_R}\right)K_I^0(t, l(t), 0) = \gamma \tag{11.1}$$

where $K_I^0(t, l(t), 0)$ is the dynamic stress intensity factor at time t for a stationary crack of length $l(t)$. Regardless of the nature of the fracture energy γ the continuum limit for the crack speed is the Rayleigh wave speed since the left-hand side of the equation, representing the dynamic energy release rate, would approach zero as this speed is approached. Long before the establishment of the above equation, a remarkable experimental observation was made by Schardin and Struth (1938): cracks set in motion by rapid impact loading quickly attained a maximum speed that was characteristic of the material, but significantly smaller than the Rayleigh wave speed; they showed this conclusively by taking time-resolved photographs of cracks propagating in inorganic glasses with a Cranz–Schardin multiple spark camera. This result has since been reinforced through hundreds of measurements by numerous research groups working on different materials. A survey of measured limiting crack speeds for nominally brittle materials is shown in Table 11.1 for crack growth in noncrystalline materials and in Table 11.2 for cracks trapped on cleavage planes in crystalline materials or along weakened planes in amorphous materials. From these results, it is clear that cracks growing on cleavage planes of crystals grow at speeds that are close to the Rayleigh wave speed. The fracture energy γ is associated primarily with the surface energy and this is expected to be independent of the speed of generation of the surface. Thus γ is constant and the crack grows at the limit set by the continuum energy release rate. In contrast, for the noncrystalline materials, the crack has to 'find' its way through the disordered material; the experimental conclusion is that the limiting crack speed is significantly lower than the Rayleigh wave speed, lying in the range of $0.4-0.7C_R$. In fact, Schardin (1959) performed an illuminating investigation: he measured the limiting crack speed in 29 inorganic glasses obtained by systematically varying the composition. The limiting crack speeds measured in these glasses are shown in Fig. 11.1. This investigation showed that the limiting crack speed, while a constant for each material, was not the same fraction of the characteristic wave speeds for all the materials examined; the liming speed varied in the range from 0.347 to $0.614C_s$. This suggests that the continuum formulation (outer problem) provides

Table 11.1 Limiting crack speeds for noncrystalline materials

Material	Author	ν	v/C_d	v/C_s	v/C_0	v/C_R
Glass	Bowden et al. (1967)	0.22	0.27	0.42	0.29	0.51
	Edgerton and Bartow (1941)	0.22	0.26	0.43	0.28	0.47
	Schardin and Struth (1938)	0.22	0.28	0.47	0.30	0.52
	Anthony et al. (1970)	0.22	0.36	0.6	0.39	0.66
Plexiglas	Cotterell (1965)	0.35	0.26	0.54	0.33	0.58
	Paxson and Lucas (1973)	0.35	0.28	0.58	0.36	0.62
	Dulaney and Brace (1960)	0.35	0.28	0.58	0.36	0.62
	Fineberg et al. (1992)					0.58–0.62
Homalite-100	Beebe (1966)	0.31	0.16	0.31	0.19	0.33
	Kobayashi and Mall (1978)	0.345	0.17	0.35	0.22	0.37
	Dally (1979)	0.31	0.22	0.35	0.24	0.38
	Ravi-Chandar and Knauss (1984a–d)	0.31	0.24	0.41		0.45
	Hauch and Marder (1998)					0.37

no clues to identification of the limiting speed and that one must look towards models of the fracture process (inner problem) in order to determine the velocity dependence of the fracture energy. Schardin suggested that the limiting crack speed be considered a new physical constant, perhaps related to other physical parameters that govern the fracture process.

It is important to state the dilemma that faces us: in nominally brittle amorphous solids, and in the absence of any rate-dependent dissipative processes, γ should be independent of the crack speed, v; why then does the crack not accelerate to the Rayleigh wave speed? Early suggestions that the speed was limited due to the onset of crack branching (discussed in Section 11.3) beg the question because branching was itself not well understood; Cotterell (1965) suggested that the maximum speed must depend on the material properties even though continuum analysis sets an upper limit of the Rayleigh wave speed. Based on an extensive investigation of the microscale aspects of fracture, Ravi-Chandar and Knauss (1984c) suggested that even in nominally brittle solids, fracture proceeds with a significant process zone in which nucleation, growth and coalescence of microcracks occur; they suggested that the dynamics of evolution of these processes and the microscopic path instabilities that this triggers provide a rate and state-dependent character to the fracture energy; thus, $\gamma = \gamma(v, K_I(t))$. The microstructural details of the fracture process are discussed in Section 11.2. More recently, Fineberg et al. (1991, 1992) observed that cracks growing faster than about $0.36C_R$ in polymethylmethacrylate

Table 11.2 Limiting crack speeds for crystalline materials and trapped cracks

Material	Author	ν	v/C_s	v/C_R
LiF	Gilman et al. (1958)			0.63
MgO	Field (1971)			0.88
Silicon	Cramer et al. (1997)		0.75	
Weak bond in PMMA	Washabaugh and Knauss (1993)			0.90

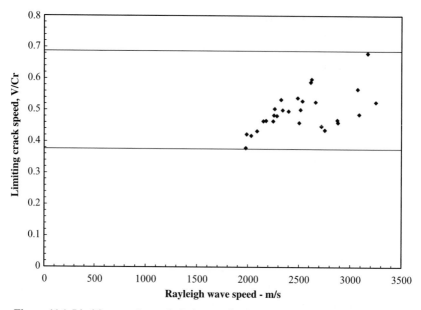

Figure 11.1 Limiting crack speeds in inorganic glasses. (Data from Schardin, 1959.)

(PMMA) exhibited rapid oscillations in the crack velocity, triggered primarily by small microbranches that are issued off the main crack; they suggested that this dynamic crack path instability was the reason for the observed limiting speed. Strong evidence that the process zone was responsible for determining the limiting crack speed was provided by Washabaugh and Knauss (1994); they prepared a specimen by diffusion bonding two plates of PMMA, but controlling the bonding process to yield a bonding plane with very low toughness. Essentially, the fracture process zone was confined to the width of the diffusion bond layer; cracks in this specimen were found to travel at about $0.90C_R$. Contrasting this with a limiting speed of $0.60C_R$ in a completely bonded specimen, it is apparent that the fracture process dictates the limiting speed. Hull (1997a,b) has expanded on these explanations by suggesting that twisting and tilting of the crack front due to microscale variations in the symmetries results in a structure to the process zone in thermoset plastics that may not be able to nucleate independent microcracks. All these models of the crack growth process are supported through different experimental measurements and observations and we shall review them in Section 11.2.

The resolution of our dilemma is, in fact, physically complete: while the continuum analysis dictates that the only way a crack can respond to additional influx of energy is by accelerating until it eventually outruns the energy at the Rayleigh wave speed, the fact that a fracture process zone has structure and dynamics associated with its evolution— regardless of the particular model—provides for rate and state dependence for the fracture energy and presents other possibilities for explaining the lower observed limiting speeds. Models of the fracture process range from idealized atomistic and lattice models to phenomenological models to mechanism-based nucleation and growth models; while none of these models have advanced to the stage of providing quantitative explanation of the limiting speed or for the dependence of the fracture energy on the crack speed, there are qualitative features of the models that appear promising. We shall first examine the overall development of crack surface roughness and the mechanisms of fracture in this chapter and then examine specific models of dynamic fracture in Chapter 12.

11.2 Fracture Surface Roughness

The fast fracture surfaces in nominally brittle materials clearly indicate a varying surface roughness—this is evident in naked-eye observations of the fracture surface. However, this is not always the case; gem cutters would be out of business were it not for the fact that in crystalline materials, fast fracture can run along cleavage planes leaving— if not atomically flat—at least optically flat surfaces. In addition, experimental measurements indicate that a crack speed approaching the Rayleigh surface wave speed is observed in crystalline materials (see Bowden et al., 1967; Field, 1971). In this case, dynamic fracture develops in such a manner that the continuum theory of dynamic fracture is appropriate. Clearly, continuum wave propagation sets the limiting speed and the crack tip fracture processes develop in a very small spatial domain at much faster rates and are not the rate limiting processes in governing the crack speed. On the other hand, in nominally brittle glasses that have been the focus of much of dynamic fracture investigations, the fracture processes that occur over a spatial domain comparable to the surface roughness dominate the dynamics of crack growth. It is the rate of development of these fracture processes that appears to set the limiting speed rather than the rate of delivery of elastic energy through stress waves.

Typical characterization of the fracture surface roughness is through its appearance to the unaided eye under normal lighting conditions; the fracture surface of the specimen from the experiment described in Fig. 8.7 is shown in Fig. 11.2. Three distinct zones are usually identified on the fracture surface and labeled as 'mirror', 'mist', and 'hackle'. In fact, in the literature on fractography of glass, the so-called mirror constant $M = \sigma_f \sqrt{l}$ is often defined where σ_f is the macroscopic stress at fracture and l is the length of the mirror zone; clearly this can be translated in terms of a stress intensity factor criterion. However, these demarcations are really an artifact introduced by the observing tool; in the mirror region the scale of roughness is small compared to the wavelength of light and as a result the fracture surface reflects light specularly appearing mirror-like. Ravi-Chandar and Knauss (1984b) measured the surface roughness in a Homalite-100 fracture surface (also shown in Fig. 11.2) and found that the roughness varied almost continuously along the crack path and that there was nothing special about the mirror–mist–hackle boundaries other than that these can be discriminated based on a visual observation. A similar variation in roughness was found by Hull (1997a,b) in Araldite, a thermosetting epoxy. Furthermore, the crack surface roughness was found to be independent of the crack speed—the speed was constant over the segment of the fracture surface shown—and depended on the stress intensity factor. Thus, the mirror-like surface appearance definitely does not indicate that the fracture surface in this region is smooth; only that the roughness has not been resolved in the observation. Higher resolution

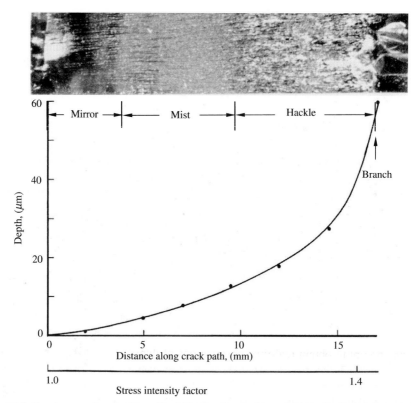

Figure 11.2 Fracture surface in Homalite-100. Crack speed was 0.38C_R. Typical 'mirror', 'mist' and 'hackle' regions are identified. The graph indicates the maximum depth of the fracture surface as a function of crack position and also the stress intensity factor. (From Ravi-Chandar and Knauss, 1984b.)

measurements with an atomic force microscope yield morphological details of the surface roughness at this scale as well. The topological as well as morphological features on the fracture surface present evidence of the events that occurred during crack propagation. Many different interpretations of these features have been presented in the literature; Ravi-Chandar and Knauss (1984b) and Ravi-Chandar and Yang (1997) present a picture that is based on nucleation, growth and coalescence of microcracks in the fracture process zone. Fineberg and Marder (1999) consider a dynamical instability that is triggered at a critical crack speed to contribute to the development of surface roughness. Hull (1997a,b) also suggested that local microbranching and tilting of microbranches results in surface roughening. It should be noted that these interpretations are not mutually exclusive, but may in fact be different interpretations of the same or similar events.

Mandelbrot et al. (1984) showed that fracture surfaces have a fractal character. Bouchaud (1997) has studied the fracture surfaces of different materials and suggested that the roughness scales in a self-affine manner; thus the fracture surface structures are invariant under an affine transformation

$$\{x_1, x_2, x_3\} \rightarrow \{bx_1, bx_2, b^{\zeta}x_3\} \qquad (11.2)$$

where b is a constant and ζ is called the roughness index or Hurst exponent. This affine scaling implies that the height at any point is given by the following:

$$h(x_1, x_2) \sim \left(\sqrt{x_1^2 + x_2^2}\right)^{\zeta} \qquad (11.3)$$

Bouchaud showed that under both quasi-static and dynamic crack growth conditions the exponent ζ was around 0.8 whenever the scale of the measurement of roughness was greater than a material-dependent length ξ_c. This exponent appears to be universal in the sense that it is independent of material, and the conditions that generated the fracture surface. For roughness at a scale below ξ_c, the roughness index was about 0.5. ξ_c is identified as a cross-over length scale at which mechanisms of roughness generation changes; it has been suggested that this crossover length scale depends on the largest heterogeneity in the material. More work remains to be done in this area to relate the roughness index to fracture energy and processes as well as to evaluate the influence of crack speed.

11.2.1 Real-Time Observations of Multiple Crack Fronts

In an attempt to obtain real-time observations of dynamic crack growth mechanisms, Ravi-Chandar and Knauss (1984b) used the repeatability of the electromagnetic loading scheme to generate high-speed photomicrographs of the crack front. These photographs were obtained in an experiment with Homalite-100 specimens repeated three times with the same loading, but with the center of the camera's field of view located in the mirror, mist and hackle regions, respectively, in each of the experiments. In the mirror zone, from Fig. 11.3a, one notes that the crack front exhibits a thumb-nail shape reminiscent of quasi-static crack propagation; this crack front curvature is due to a change from a plane strain to plane stress constraint. The crack front in the middle of the specimen leads the crack front at the plate ends by about 0.5 mm and there is a smooth variation in the crack front between the middle and face of the plate. At the faces of the plate the crack front forms a caustic and this leads to the crack front appearing to be wider than the plate width (see Section 8.3 for a discussion of caustics and Ravi-Chandar and Knauss (1984b) for a discussion of this particular experimental

Figure 11.3 High-speed photomicrographs of a moving crack front in Homalite-100. Crack front is located in the (a) 'mirror' zone; (b) 'mist' zone and (c) 'hackle' zone. Photographs were obtained with the crack front observed obliquely. (Reproduced from Ravi-Chandar and Knauss, 1984b.)

arrangement). Fig. 11.3b shows the crack front in the mist zone and it appears distinctly different from that in the mirror zone. Again, caustics are formed at the faces of the plate and these are larger because of the higher stress intensity factors associated with the mist zone. The most striking difference is that the 'crack front' is nearly straight but exhibits a number of small caustics. This indicates that there is no longer a single crack front, but an ensemble of cracks propagating along the apparent crack front and generating multiple caustics. Note that this image may also be interpreted as a single crack front splitting into multiple fronts as suggested by Hull (1997a,b), and not necessarily into multiple microcracks. The image in Fig. 11.3c shows the appearance of the crack front in the hackle zone which seems to be similar to that in the mist zone but represents the coarser structure of the fracture surface in the process of forming. This might have been expected in view of the fact that post-mortem examinations of the mist and hackle surfaces exhibit similarities. However, the caustics appear larger indicating still higher loading; furthermore, along the crack front fewer but larger caustics are observed. Microbranches are also visible in these photographs; clearly these frustrated branches do not develop through the entire thickness of the specimen. From these photographs and from post-mortem examination of the corresponding fracture surfaces, Ravi-Chandar and Knauss (1984b) suggested the following picture of dynamic crack growth: initially, in the mirror zone a 'single crack' propagates with a curved crack front similar to that observed in quasi-static crack propagation. In the mist zone several small cracks propagate simultaneously and the ensemble crack front is nearly straight. Further crack propagation is really governed by the details of the interaction between these microcracks. In the hackle zone crack growth occurs by the same physical process, except that the size scale of the microfracturing increases. Real-time photomicrographs of a crack branching event were also captured; these are described in Section 11.3.

11.2.2 Fast Fracture Surfaces in Polymethylmethacrylate

A number of investigators have reported observations of parabolic markings (Smekal, 1953; Irwin and Kies, 1952; Leeuwerik, 1962) and a periodic morphology in the dynamic fracture surface of PMMA (Green and Pratt, 1974; Döll, 1976a,b; Fineberg et al., 1991; Washabaugh and Knauss, 1993). Fineberg et al. (1991) suggest that the periodicity appears at a critical velocity and that the surface roughness is well correlated with measured velocity fluctuations, while Washabaugh and Knauss (1993) indicate that the periodicity correlates well with the stress intensity factor. Fineberg et al. (1991) also suggest that this periodic phenomenon is responsible for limiting crack speeds to about 50% of the theoretical limit of the Rayleigh surface wave speed. Here, we first describe the experimental observations of Fineberg et al. and then follow this with fractographical examinations of Ravi-Chandar and Yang (1997).

Fineberg et al. (1991) performed dynamic fracture experiments in a narrow strip configuration shown in Fig. 11.4a. Typical specimen dimensions were 100–200 mm wide and 200–250 mm tall, with thickness of either 1.6 or 3.2 mm. The initial crack length was only about 3–4 mm. Load was applied by moving the top grip in steps of 10^{-4} of the height and holding for 10–20 s in order to allow crack initiation with very

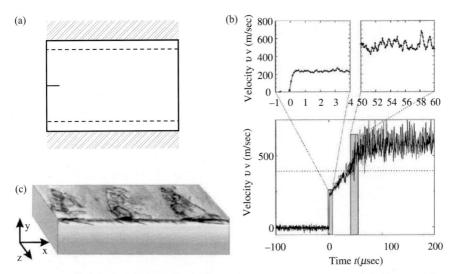

Figure 11.4 (a) Strip specimen geometry; length ~250 mm, width ~200 mm, initial crack length ~3–4 mm. (b) Crack speed measured with the potential drop technique. The lower curve shows the overall time history while the upper curves show details near the initiation time and the time at onset of dynamic instabilities. (c) Fracture surface of the PMMA specimen in the region of crack speed oscillations. *x–z* is the nominal plane of the fracture surface; significant roughness and periodic bands can be seen. (Reproduced from Fineberg and Marder, 1999.)

little time dependence in the loading. In this configuration, the energy release rate remains constant at a value that can be determined from the load measured at the onset of fracture propagation. Fineberg et al. used a potential drop method to determine the crack position to an accuracy of about 0.2 mm and crack speed to an accuracy of about ± 5 m/s. A typical measurement of the crack speed as a function of time from one of their experiments is shown in Fig. 11.4b. In these experiments they found that the crack speed typically jumped to a value of about 200 m/s within 1 μs. Then, the crack accelerated smoothly until reaching a critical speed v_c; beyond v_c large oscillations in the crack speed were observed. Through observations of the fracture surface features, these crack speed oscillations were associated with a microbranching instability; the structure of the fracture surface is shown in Fig. 11.4c. Fineberg et al. (1991) have suggested that this microbranching instability is responsible for large energy dissipation (through the creation of extra surface area in the microbranches). In fact, through careful measurements of the profile and the density of the microbranches, Sharon and Fineberg (1996) have correlated the increase in area generated by the microbranches to the increase in fracture energy. Questions regarding the origin of the crack path instability (microbranches) and the periodic band formation remain to be answered: What are the micromechanisms governing the fast fracture of PMMA? What triggers the periodicity in the fracture surface morphology? We discuss the fracture mechanisms next.

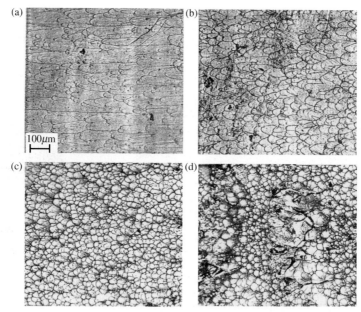

Figure 11.5 Fracture surface in a PMMA specimen. The density of conic markings increases from about 350 mm² in (a) to about 2600 mm² in (c) and (d). The periodic bands seen in Fig. 11.4c can be seen in (d) at a larger magnification. (Reproduced from Ravi-Chandar and Yang, 1997.)

An examination of the fracture surface of PMMA is shown in Fig. 11.5. The surface is tiled with a pattern of conic[1] marks (Fig. 11.5a); these markings are well known in the fracture of amorphous materials. They are usually interpreted as being level differences resulting from an encounter between a microcrack and a main crack. Along the crack path, the number and density of the conic marks increase (see Fig. 11.5b–d). In addition, their eccentricity appears to change perhaps indicating a difference in the velocity of approach between the microcrack and the main crack(s) or equivalently the ratio of the nucleation distance and the nuclei spacing. At higher load levels, a version of the periodic morphology observed by Fineberg and Marder (1999) appears to emerge from this state of increasing density of conic marks; this is clearly observable in Fig. 11.5d. The spacing, size and depth of these periodic bands also evolve along the crack path—the period of these bands increases from about 300 μm to about 500 μm; in fact, the periodic bands begin independently at several different sites across the plate thickness as very small spots and then gradually grow in width as the crack propagates, indicating significant three-dimensionality in its evolution. Washabaugh and Knauss (1993) observed that coincident with the appearance of the periodic surface roughness, stress waves are visible in high-speed photographs; they associated these periodic bands with intermittent growth of the crack.

---

[1] In the literature, these are referred to as parabolic markings. In general they are conic markings; some are parabolic while most are hyperbolic. Some elliptic marks have also been identified. We will refer to these collectively as conic markings.

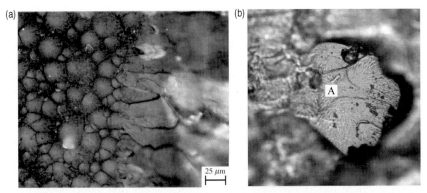

Figure 11.6 Fracture surface in a PMMA specimen. (a) The region just ahead of periodic bands seen in Fig. 11.5d; note that the conic marks are terminated abruptly. (b) A close-up view of the rough regions inside the periodic bands showing multiple microcracking. 'A' is the first microcrack that was nucleated by the 'main crack'; other microcracks were nucleated by this microcrack as it grew. This is evident from the fact that all the conic marks appear to focus at the point A. (Reproduced from Ravi-Chandar and Yang, 1997.)

The fracture surface shown in Fig. 11.5 indicates that *the periodicity of the fracture surface is not really due to a change in the mechanism of crack growth*: dynamic crack growth is always governed by microcrack growth and coalescence. The flat surface tiled with conic marks of different sizes is generated by a continuous (perhaps) sequential nucleation, growth and coalescence of microcracks whereas in the periodic bands a rather rough surface is created because microcracks or microcrack clusters formed far ahead of the main ensemble crack coalesce with it, with surface asperities in the range 20–30 μm. In fact, the conic marks in the flat region before the band end abruptly and give way to the rough surface as can be seen from the high-magnification photograph shown in Fig. 11.6a. In Fig. 11.6b, the surface of one of the periodic bands is shown: that this is at a different plane from the 'main' crack plane is obvious from the defocusing of other areas in the optical micrograph. Clearly, the surface of these bands is also tiled with conic marks. Notice that the flaw identified by A in this figure nucleated first by the 'main crack' and its growth has led to the nucleation of the other microcracks in this surface. This cluster of microcracks develops in the enhanced stress field of the main crack, but is quite independent of the main crack. The two cracks merge by breaking the ligament if the conditions are favorable. This process of forming off-axis microcrack clusters and coalescing them with the main crack could be repeated leading to the periodic banding; on the other hand, if the load levels are sufficiently high, the off-axis microcrack clusters could outrun the main crack and become full-fledged crack branches. While the above discussion gives compelling evidence of brittle microcrack-based fracture mechanism that results in the periodic morphology and crack branching, it is insufficient to identify the period; this will have to wait until the appropriate crack growth equations for the microcrack model of crack propagation are identified.

11.2.3 Origin of the Microcracks

Smekal (1953) postulated a very simple mechanism for the formation of the conic markings on the fracture surface. His model is simply that in the enhanced stress field of a primary crack, inhomogeneities triggered the initiation of a secondary crack ahead of the primary crack front; the secondary fracture may not be in the same plane as the primary front and when these two fronts intersect in space and time, the ligament separating the two cracks breaks up leaving a conic marking on the fracture surface. The conic marking thus indicates a level difference boundary, marking the common space–time interaction of the two fracture fronts with the focus of the conic identifying the origin of the secondary fracture front. Given that such conic markings are observed in a wide range of materials with very different microstructures, the microcrack interaction model appears to be the most appropriate characterization. There is, however, some disagreement over the origin of the secondary microcracks: what is the nucleus of the microcracks? Irwin and Kies (1952) pointed to small cavities or inhomogeneities at the focus of the conic markings in cellulose acetate and steels. Leeuwerik (1962) observed a bowl shaped cavity with a diameter of 0.3 μm at the foci of the conic markings. All these authors speculated that these are voids (flaws) that are naturally randomly distributed throughout the material. Cotterell (1968) suggested that perhaps these voids are formed inside the crack tip craze region. Matsushinge et al. (1984) examined the formation of the secondary cracks using acoustic emission and *post-mortem* X-ray microanalysis. While contamination in the form of Si was found in a small percentage of the markings that were examined, they could not conclude that such foreign objects were the nuclei for all the secondary fractures. It is quite possible that randomly distributed flaws acting as the origin for the microcracks is the appropriate model; it is also possible that voids are nucleated in the high-triaxial tensile field near the crack tip. A simple estimate of the stress field in which the microcracks nucleate may be obtained by considering the K-dominant field near the crack tip.

For the specimen shown in Fig. 11.4, the dynamic stress intensity factor at crack initiation was 1.03 MPa$\sqrt{m}$ and gradually increased to about 1.2 MPa$\sqrt{m}$. The yield tress (determined from quasi-static experiments) for PMMA is approximately 100 MPa and is likely to be higher at the high strain rates encountered at the dynamically growing crack tip. In an effort to estimate the size of the plastic zone, let us assume that PMMA obeys a power law hardening behavior of the type $\varepsilon = \sigma^n$, with $n = 3$. The size of the plastic zone[2] under plane strain conditions that are appropriate to the interior of the specimen is then estimated to vary from about 2.8 to 3.8 μm as the stress intensity factor increases from 1.03 to 1.2 MPa$\sqrt{m}$. Note that the plastic zone is very small, and in particular smaller than the average nucleation distance (see Section 11.2.5 for the definition of nucleation distance); thus conditions of small-scale yielding seem to be appropriate. Assuming then that the linear elastodynamic stress field estimates are appropriate at these distances, the stress-state at the nucleation of the flaws can be determined. The triaxial stress state at the nucleation distance of 5.5 μm, corresponding to a stress intensity factor of 1.03 MPa$\sqrt{m}$

---

[2] This is a quasi-static estimate; under dynamic conditions, since the yield stress is likely to be higher, and also due to the shortening of the zone caused by inertia effects, the plastic zone will typically be smaller than this estimate. Hence we have an upper bound estimate of the plastic zone size.

and a crack velocity of about $0.25C_R$ can then be calculated as: $\sigma_{11} = 189$ MPa, $\sigma_{22} = 175$ MPa and $\sigma_{33} = 125$ MPa; similarly at the nucleation distance of 8.5 μm, corresponding to a stress intensity factor of 1.2 MPa$\sqrt{m}$, and a velocity of about $0.25C_R$, the triaxial stress state is given by: $\sigma_{11} = 172$ MPa, $\sigma_{22} = 164$ MPa and $\sigma_{33} = 118$ MPa; note that these stresses are much higher than typical values quoted for nucleation of crazing in PMMA under uniaxial tension (Argon and Salama, 1977), which is in the range of 60–100 MPa. It appears that at such high-triaxial tensile stress levels, extensive cavitation occurs and that these cavities are the nuclei for further development of the dynamic fracture process; in this sense of cavitation, these are stress-induced or stress-activated flaws or nuclei and hence the increase in the number density of microcracks with the crack tip stress level. Of course, these cavities will develop preferentially at sites where inhomogeneities in the nanometer scale exist and facilitate the initiation of the cavities and hence the randomness in the spatial distribution of flaws on the surface.

11.2.4 Geometry of the Conic Markings

By applying Smekal's model to the conic markings in the micrographs, we can obtain information concerning the size, shape (eccentricity), and critical distance of the secondary microcracks. The interaction of straight fronted main crack with a radially growing microcrack and the further interaction of this microcrack with another microcrack is depicted in Fig. 11.7a. Consider a planar crack front approaching a microcrack nucleus with a velocity v_c; when the distance between the crack front and the nucleus is d_n, the microcrack nucleates (i.e. it begins to grow) and grows radially symmetrically at a velocity v_{c1}; d_n is the critical nucleation distance and is presumably dictated by the inherent characteristics of the material and the stress field. A second nucleus is at a spacing s from the first nucleus; the spacing s will in general follow some statistical distribution, perhaps dictated by the local stress field and the material microstructure; when the distance between the growing front of the first microcrack and the second nucleus is equal to d_n— the nucleation distance—the second nucleus begins to grow radially symmetrically with a velocity v_{c2}. The microcrack nuclei are usually not on the same plane as the main crack, but offset by about a few microns; thus the interaction between the main crack and the microcracks leaves a trail of their common space–time interaction. If the spacing, s, the nucleation distance, d_n, and the velocities, v_c, v_{c1}, and v_{c2}, are known, the resulting conic marking on the fracture surface may be calculated as indicated by the solid line in Fig. 11.7a. Such an interaction of a planar crack front with two microcrack nuclei is observed in experiments as shown in Fig. 11.7b. Alternatively, from the micrographs, one might extract details regarding s, d_n, v_c, v_{c1}, and v_{c2}. For simplicity, assume that the velocities of the main crack and microcracks are identical and equal to v. Then the equation describing conic marking on the fracture surface is given by

$$\frac{(2x_1 + s - d_n)^2}{(s - d_n)^2} - \frac{4x_2^2}{(2s - d_n)d_n} = 1 \qquad (11.4)$$

By comparing the measured conic with Eq. 11.4, one can either verify that the microcracks grow with the same speed or determine the appropriate speed ratio. Note that assuming

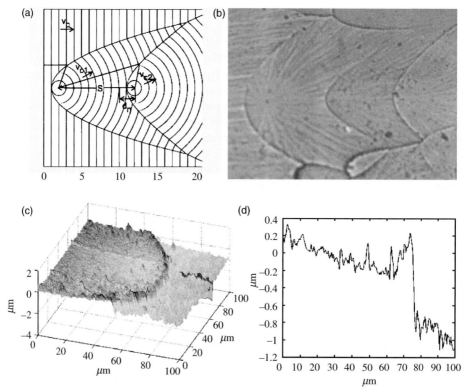

Figure 11.7 **(a) Geometry of formation of conic markings on the fracture surface due to nucleation and growth of microcracks. (b) Experimentally observed conic marks on the fracture surface of PMMA. (c) AFM image of the topography of a conic mark. (d) Topography of the conic mark along a radial line. ((a) and (b) reproduced from Ravi-Chandar and Yang, 1997.)**

the velocities of the two microcracks to be equal leads to a shape of the conic that is independent of the velocity; Kies et al. (1950) demonstrated this very nicely by performing a displacement-controlled experiment where the crack propagated only a very short distance at each displacement increment. However, the growth was dynamic, and secondary cracks were nucleated ahead of the main crack; conic marks were observed on the fracture surface, decorated with the arrest lines corresponding to each displacement increment. Thus it appears reasonable to assume that the microcracks grow at the same speed; the value of this speed must still be controlled by the applied stress field and the local stress state in the fracture process zone. The variation of height in a conic marking is shown in Fig. 11.7c; a trace along one radial line of the conic marking is shown in Fig. 11.7d. Clear level differences across the conic marking of the order of a micron can be observed. Note the contrast between this level difference and the difference between the different planes when the periodic features appear on the fracture surface. Also, the boundary between the two microcracks is seen to have been stretched during the cracking process.

11.2.5 Statistics of Microcracks in PMMA

The contours of a large number of conic markings were digitized and then fit with a general second-order equation to determine both the eccentricity and the distance to the focus. The eccentricity values were close to unity in region corresponding to Fig. 11.5a but were slightly higher in subsequent regions (Fig. 11.5b–d); assuming that the velocities of the microcracks are all equal, the secondary microcracks were nucleated by nearly planar crack fronts in the early stages of crack growth, but in later stages, since the spacing between the microcrack nuclei is small, nucleation of microcracks is triggered by the curved crack fronts of other microcracks. This is supported by the fact that the average spacing between nuclei decreases from about 53 μm in Fig. 11.5a to 33 μm in Fig. 11.5b and 20 μm in Fig. 11.5c. The nucleation distance can be obtained from measurements of the distance to the foci; the average focal distance increases from about 2.75 μm in Fig. 11.5a to 3.75 μm in Fig. 11.5b to 4.25 μm in Fig. 11.5c; this indicates that the nucleation distance d_n increases from about 5.5 μm in Fig. 11.5a to 7.5 μm in Fig. 11.5b to 8.5 μm in Fig. 11.5c. It is clear that the conic markings on the fracture surface indicate *an increase in the number of nuclei activated into growing along the crack path and an increase in the nucleation distance at which the secondary microcracks begin to grow*. Since the average crack speed is constant in these regions, the increasing stress intensity factor (actually the increasing near tip stress field) is the driving force in generating this large density of microcracks. Thus, as the stress intensity factor increases, more flaws are nucleated and they are nucleated farther ahead of the main crack; *this appears to be the primary mechanism of crack growth in PMMA*. Note that this is the case regardless of whether one observes steady-state crack growth, periodic banding and crack branching.

A number of investigators have examined the population of the microcracks and attempted to correlate them to the velocity and/or the stress intensity factor or fracture toughness (Cotterell, 1968; Carlsson et al., 1973; Matsushinge et al., 1984). From Fig. 11.4, we can infer that while the average crack velocity is nearly constant, and the stress intensity factor increases slowly, the density of conic markings increase dramatically; thus, the density of conic marking should be correlated to the stress or the stress intensity factor rather than the velocity. We note that the density of conic markings observed by the investigators listed above is on the order of $1-50$ per mm^2; in contrast, the density of conic markings in Fig. 11.5 is $1-2$ orders of magnitude higher! This could not only be due to differences in the molecular weight of the PMMA, but also perhaps due to differences in the stress level applied and the transient nature of the applied loading.

To interpret the conic markings in Fig. 11.5a–c quantitatively, the number of conic markings was counted and the density was determined. This was achieved by breaking up each micrograph into six regions and counting the number density of conic markings in each region. The results can be summarized as follows: first, the density of conic markings in Fig. 11.5a is about 350 per mm^2, an order of magnitude larger than in the observations in the references cited above. Secondly, there is an increase in the density of conic markings from Fig. 11.5b (900 per mm^2) to Fig. 11.5c ($1500-2600$ per mm^2); along the crack path the stress intensity factor increases at nearly constant average crack speed, suggesting that the stress-induced nucleation of secondary microcracks is the appropriate mechanism. Finally, in Fig. 11.5c, a rapid increase in density of conic markings is experienced just prior to the onset of the periodic banding morphology, indicating that a large density of flaws activated into nucleation might be the trigger for the periodic banding.

11.2.6 Growth of Microcracks

The growth rate of the microcracks cannot be easily characterized from post-mortem examinations of the fracture surface. While a description of the geometry of the conic markings can be provided if the spacing between nuclei, the nucleation distance and the respective velocities of the microcracks are given, the inverse problem is not unique: given the spacing, nucleation distance and the form of the conic marking, it is not possible to determine the velocity uniquely, especially if there is significant variation. The simplest approach is to assume that the microcracks grow with a constant velocity, but this may not be appropriate when the microcrack density becomes large. Also, we have interpreted fracture surface markings as being the result of a sequential nucleation and growth of flaws; it is conceivable that many nuclei ahead of the crack may be activated simultaneously and this possibility needs to be investigated further. Indeed, as remarked earlier, the nucleation, growth and coalescence of microcrack clusters away from the main crack tip is the primary cause for the rough crack surface, periodic banding and crack branching. Given the criterion for nucleation and growth of microcracks, the microcrack-dominated dynamic crack growth model that is described above can be simulated numerically to reveal the features observed in experiments. A simple implementation of this model was descried by Ravi-Chandar and Yang (1997). A damage model that mimics the loss of stiffness of microcracked materials was developed by Johnson (1992) and more recently by Klein and Gao (2001). These models, discussed in Chapter 12, appear to do well in predicting the experimental observations.

11.2.7 Solithane 113

Solithane 113 is a polyurethane elastomer manufactured by Thiokol Corporation. The glass transition temperature is around $-20°C$ and hence the tests on Solithane 113 were performed at $-90°C$, well into the glassy regime; thus, this material also exhibits a brittle fracture response. The crack speed in these experiments was around 400 m/s. Precise measurements of the Rayleigh wave speed in this material at $-90°C$ are not available, but one might expect that the crack speed is roughly the same fraction of the Rayleigh wave speed as in other materials. Fig. 11.8 shows the fracture surface in Solithane. The similarity to PMMA, in spite of the microstructural differences, in all aspects of the fracture surface morphology, is complete and striking. This similarity suggests the possibility that a microcrack-based model might be capable of capturing the behavior of many different brittle materials.

11.2.8 Polycarbonate

Polycarbonate (PC) is a noncrosslinked thermoplastic polymer; it is capable of significant inelastic deformation due to the relatively high mobility of the carbonate segments of its structure. However, at high rates of loading, it does exhibit brittle dynamic

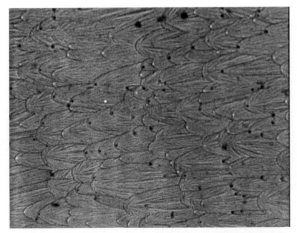

Figure 11.8 Fracture surface in the elastomer Solithane 113 indicating tiling with conic marks. (Reproduced from Ravi-Chandar and Yang, 1997.)

fracture and this is the regime considered here. The fracture surface from a polycarbonate specimen, loaded by a projectile impact as in the experiments of Taudou et al. (1992) is shown in Fig. 11.9. The corresponding crack velocity and stress intensity factor were again monitored by high-speed photography; the crack propagated at about 480 m/s which is roughly 50% of the Rayleigh wave speed, similar to the observed limiting speed in other materials. The stress intensity factor, measured using the method of caustics, increased only slightly from the value at initiation ($K_{IC} \sim 4.9$ MPa$\sqrt{m}$). The fracture surface exhibits changes in its morphology that are very different from those seen in PMMA. In Fig. 11.9a, during the early stages of crack growth, the surface is remarkably uniform; conic markings typical of PMMA and Solithane 113 are not seen at all. Further along the crack path, a banded morphology develops as shown in Fig. 11.9b, but the bands are very shallow, and their period is only about 60 μm; these appear to be at about the same scale as the conic marks in PMMA, but apparently are not generated by microcracking. Doyle (1983) observed a similar banded structure in polystyrene (PS) at 80°C with roughly the same morphology and spacing; he interpreted these as being formed by a dominant craze at the crack tip. As the crack extends, the banded morphology is interspersed with another periodic structure with a period of roughly 300 μm and a larger surface roughness, as shown in Fig. 11.9c. The roughness of the fracture surface along the crack path and the size scale of the periodicity appears to increase. Although not clearly visible in Fig. 11.9c, contained within each 300 μm band are 60 μm bands similar to the ones observed in Fig. 11.9b. Remarkably, the period of this large periodic surface feature is about the same as the periodicity in PMMA and Solithane 113; Doyle (1983) also observed this larger periodic banding with a spacing again in the range of 300 μm in polystyrene at room temperature, and attributed it to multiple crazes being generated at the crack tip due to rapid stress build up under the dynamic loading. The larger periodic surface features in Fig. 11.9c are clearly a manifestation of level changes in the fracture surface; even though clear evidence of microcracking (through the conic markings) is not seen on the fracture

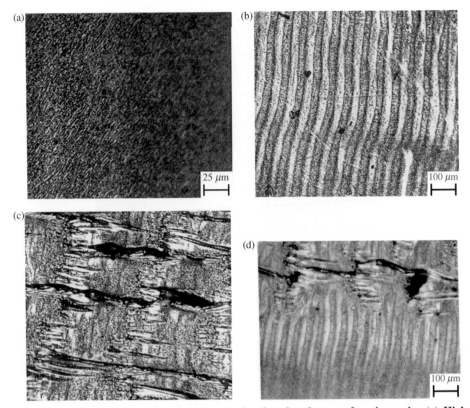

Figure 11.9 Fracture surface in polycarbonate showing the absence of conic marks. (a) High-magnification image in the low stress intensity region. (b) Periodic striations with a spacing of about 60 μm. (c) Periodic bands with a spacing of 300 μm. (d) Variation in the fracture surface across the specimen width. (Reproduced from Ravi Chandar and Yang, 1997.)

surface of PC, these level differences can be interpreted as being caused by the progression of fracture caused at different levels by microcracks. If a craze is initiated first and then forms a microcrack as it breaks down, one might not observe the characteristic conic marks. Thus, one might consider that in the early stages of the crack growth process, a single craze is formed across the entire plate thickness and grows by advancing both the craze and crack tips; hence the absence of any remarkable features in Fig. 11.9a. As the stress level continues to rise, the crack tip catches up with the craze tip and a sequence of new craze development and breakdown follows leading to the formation of the 60 μm banded surface feature. As this process continues under increasing stress levels, the single craze spanning the entire width of the specimen is not stable and breaks up into many independent craze cracks along the width of the specimen. These are the level differences observed in the 300 μm bands. This model might provide a link between the fracture mechanisms in PMMA, PC and PS. We have observed under different loading conditions periodic bands similar in structure to those in Fig. 11.9b, but with periods ranging from 4 to 60 μm; these bands are still under investigation.

There appears to be a significant three-dimensional influence on the nucleation and growth of these features. Near the plate surfaces, the fracture surface always displays a boundary layer where a transition is observed from a smooth surface to a 60 μm periodic striation and to the 300 μm periodic feature as shown in Fig. 11.9d. This three-dimensional nature of the surface variation is a clear demonstration that the crack speed is not the driving factor in determining the periodicity since the speed variations along the crack front across the width are not expected to be significant; the real-time photographs of the crack fronts shown in Fig. 11.3 support this claim. However, stress variations are quite significant across the thickness due to changes in the constraint in the thickness direction from a plane strain state in the interior to a plane stress state in the boundary layer near the surface of the plate. A similar boundary layer is also observed in PMMA, where the periodic banding caused by the microcrack clusters does not develop in a thin layer near the plate surfaces and in other materials.

11.2.9 Homalite-100

Homalite-100 is a thermoset polyester that has been used in dynamic fracture studies by many investigators because of its nominally brittle behavior. The evolution of the fracture surface in this material has been considered in detail by Ravi-Chandar and Knauss (1984b). The maximum observed crack speeds in this material are again around 50% of the Rayleigh wave speed. The fracture surface morphology appears to be quite different from that in PMMA, Solithane and polycarbonate; individual microcracks identifiable by their traces on the fracture surface in the form of conic marks are only rarely seen. Also, Homalite-100 does not exhibit tractable periodicity on the fracture surface in any velocity range up to the limiting speed. The surface roughness, however, appears to evolve rather continuously along the crack path; Ravi-Chandar and Knauss (1984b,d) proposed that this was due to the evolution of the microcrack-dominated process zone. In recent experiments in the strip configuration, Hauch and Marder (1998) observed crack surface roughening similar to that shown in Fig. 11.2; in addition to the parabolic marks, attempted microbranches, they showed a periodic microbranching phenomenon similar to that shown in Fig. 11.4.

11.2.10 Other Brittle Materials

The fast fracture surface morphology in many other brittle materials exhibit similarities to the features presented here. Among amorphous materials, conic markings were observed in silicate glasses by Smekal (1953), in polystyrene by Regel (1951), and in cellulose acetate by Kies et al. (1950). Leeuwerik (1962) observed conic markings in spherulitic nylon, a semicrystalline polymer. Irwin and Kies (1952) report conic markings in polycrystalline materials such as steel, copper molybdenum and lead–tin alloys and in a large (single?) crystal of potassium bromide. Thus, it appears that the microcrack-based crack growth mechanism might be appropriate for a whole class of brittle materials. The origin, nucleation and growth mechanisms and kinetics, in each material could be quite different, but one might expect a common framework in their modeling.

Hull (1997a,b) provided a different interpretation of the evolution of the fracture surface roughness. He discounted the possibility that microcracking can occur in brittle thermosetting polymers suggesting that the flaws required to nucleate such microcracks must be very small. Instead, Hull proposed that roughening is generated by tilting of the crack front out of the plane of the main crack. The tilting was associated with nucleation of steps on the fracture surface and the associated mixed-modes I and III generated at the crack tip. This model of crack tilting is used to generate multiple crack fronts, not multiple microcracks, with each crack front developing along different planes and resulting in roughness. The level differences between the different crack fronts was shown to increase with increasing stress level and not correlated to the crack speed. While there are specific differences between this model and the microcrack model discussed above, the similarity lies in the use of multiple crack fronts or cracks associated with the break-up of a single propagating crack as the source of roughening.

11.3 Crack Branching

Branching of cracks in glass was recorded by Schardin (1959); other investigators have observed crack branching in crystalline as well as amorphous materials. Spectacular patterns are formed when dynamically growing cracks break up into multiple cracks; an example obtained in an experiment using the electromagnetic loading is shown in Fig. 11.10. Dally (1979) observed multiple branches emanating from a single crack in an explosively loaded crack in Homalite-100. Kobayashi et al. (1973) examined branching in quasi-statically loaded specimens; cracks were driven into an increasing stress field and branching was promoted. From isochromatic fringe patterns observed near the propagating and branching cracks, these authors determined the stress intensity factor and crack speed histories. They concluded that the critical stress intensity factor at

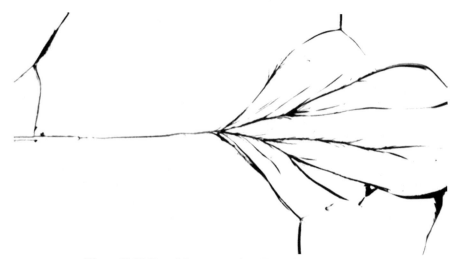

Figure 11.10 Branching pattern in a Homalite-100 specimen.

branching was between 2.4 and 3 times the initiation toughness. Congleton and Petch (1967) attempted to estimate the stress intensity factor at branching by considering a small Griffith crack—a microcrack in the crack tip process zone—placed ahead of the main crack in order to explain crack branching; this appears to be the first attempt to tie the fracture evolution in terms of microscale fracture processes. However, this phenomenon has eluded analytical description, even though the search for an explanation to branching triggered early interest in dynamic fracture analysis.

Yoffe (1951) attempted to explain the branching of cracks from an analysis of the problem of a crack of constant length that translates with a constant velocity in an unbounded medium. From this solution she found that the maximum of the hoop stress acted normal to lines that make an angle of 60° with the direction of crack propagation when the crack speed exceeded $0.60C_s$; the angular variation of $\sigma_{\theta\theta}$ is shown in Fig. 3.6. Therefore, Yoffe suggested that this stress field rearrangement might lead to crack branching. However, as was reviewed in Section 11.1, cracks seldom reach this speed, but branch nevertheless; so the macroscopic stress field rearrangement, is attractive as it is because of its simplicity, is not likely to explain macroscopic crack branching. Eshelby (1970) argued that since at least twice as much area is to be created after branching, and since the energy available cannot change discontinuously, the crack would not branch unless the factor $g(v)$ in Eq. 11.1 could be doubled with a corresponding reduction in the crack speed; while it is difficult to determine the pre-branching crack speed that would result in this doubling, estimates based on zero branching angle indicate that this speed is around $0.50C_R$. While this argument is appealing from an energetic point of view, and must be correct in principle, it is deficient in two aspects. First, there is an inherent assumption that the energy available during the pre-branching growth is sufficient to propagate only one crack; second this argument would require the branched cracks propagate with a significantly reduced speed at least in the initial stages of branching. Both are contrary to experimental observations. The process zone of the crack prior to branching is quite large, and as the crack grows at the limiting speed it dissipates many times the energy required for the growth of a crack with a mirror surface. Based on measurements, the dynamic stress intensity factor near crack tips growing at about $0.50C_R$ could be as large as three times the dynamic initiation toughness; furthermore, the dynamic stress intensity factor changes significantly for very small changes in the crack speed (see Figs. 10.19–10.24). Thus, the crack does not have to change speed significantly in order to make available additional energy for growing two or more cracks; it simply has to find a mechanism to break up into two or more cracks and can continue to grow at the same speed, but perhaps with a slightly smaller process zone. In fact, experimental measurements bear this out; as we shall see later, the crack speed does not change upon branching.

We describe here some experiments performed to examine the underlying macroscopic and microscopic aspects of crack branching. In the first experiment, a narrow strip specimen, 500 mm long and 50 mm wide and 4.76 mm thick, with a notch machined parallel to the long side of the rectangle to simulate the crack was used. The electromagnetic loading scheme described in Section 6.6 was used to generate a uniform pressure loading over the crack surfaces. Selected sequence of high-speed photographs from this test are shown in Fig. 11.11. The time variation of the dynamic stress intensity

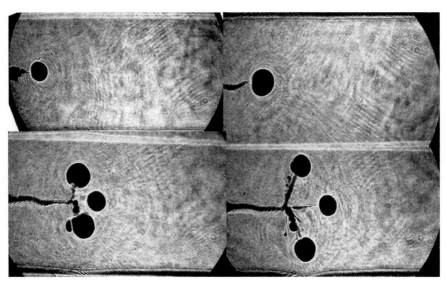

Figure 11.11 Selected frames from a high-speed sequence showing caustics at the tip of a branching crack. The vertical dimension is 25 mm.

factor measured with caustics and crack position are shown in Fig. 11.12. Crack initiation occurred at about 20 μs, as the stress intensity factor increased to about 0.45 MPa√m. The crack began to grow at a constant speed of about 430 m/s. The stress intensity factor continued to increase (caustics interpreted with the assumption of a *K*-dominant field) to about 1.2 MPa√m at about 90 μs. At this time, the crack branched into three distinct cracks, with one branch continuing along the original crack direction and the other two moving at an angle of 65° and 70° from the main crack for the top and bottom branches, respectively. Beyond branching, the three cracks continued to grow without any measurable change in the speed for the next 50 μs; however, the dynamic stress intensity factor at the continuation of the main crack dropped significantly to a value that is closer to the initiation toughness. The newly developed branches are also at a stress intensity factor that is comparable to the crack initiation level. Continued loading from the pressure loading on the parent crack and the arrival of reflected stress waves in the narrow strip configuration cause additional increase in the stress intensity factor. Examination of the fracture surface revealed that the pre-branching crack roughness was in the 'hackle' range while the fracture surfaces of the three branches were in the 'mirror' range. While in this particular example reflected waves arrived rather quickly, the only role of these waves was to generate a larger loading at the crack tip; crack branching was also seen in the experiment shown in Fig. 8.7 in the absence of reflected waves, with similar characteristics. Measurements from different investigators indicate that branching occurs when the stress intensity factor reaches a critical value that is between two and three times the quasi-static fracture toughness of the material, but the crack speed at branching varied in different experiments.

Variations in the crack branching angle were also found; for the test in Fig. 8.7, the angle included between the two branches is 62° while each branch is inclined at 65° to 70°

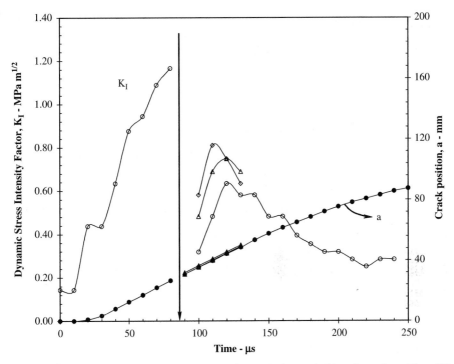

Figure 11.12 Time history of the stress intensity factor (open symbols) and crack position (filled symbols) for the experiment shown in Fig. 11.11. The vertical arrow indicates the branching point obtained from crack position measurement.

in the test in Fig. 11.11. Clearly, the macroscopic stress field also influences the angle and number of branches that appear. Ravi-Chandar and Knauss (1984d) explored this by altering the crack parallel compressive stress at the branch location; the following loading was generated. With the electromagnetic loading device, a pressure of magnitude P_1 was applied on the crack surfaces. A second loading pulse with a pressure P_2 was applied to the specimen from a second electromagnetic loading device, with the pressure parallel to the crack. This situation is illustrated in Fig. 11.13. The repeatability of the loading scheme and the electrical synchronization of the events made it possible to time the events such that the secondary stress pulse (P_2) arrived at the point of anticipated branching just prior to the instant of crack tip arrival at that point. In this arrangement, the magnitude P_2 of the secondary pulse was varied systematically to examine its effect on the macroscopic aspects branching. It was observed that depending on the magnitude of the dynamic crack parallel stress (compression), crack branching was suppressed for some time and the angle of subsequent branching was changed. Two images from such tests are shown in Fig. 11.13, showing clearly the delay in branching and the decrease in the branching angle. The marker in the photographs indicates the length at branching L_{br} for the case $P_2 = 0$. The change in L_{br} and the angle of branching θ_{br} with P_2 are shown in Fig. 11.14. In contrast, in the experiment shown in Fig. 11.11, many reflected waves bounce to the crack tip region and during the time of branching, it is expected that the crack parallel stress should be

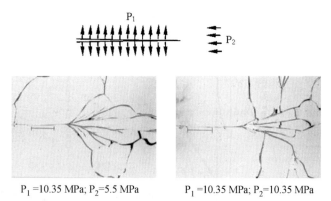

P₁ =10.35 MPa; P₂=5.5 MPa   P₁ =10.35 MPa; P₂=10.35 MPa

Figure 11.13 Assembled plates from tests where crack-parallel compressive stress wave was imposed to delay branching. The marker in the photographs indicates the length at branching in the absence of the compressive stress wave. (Reproduced from Ravi-Chandar and Knauss, 1984d.)

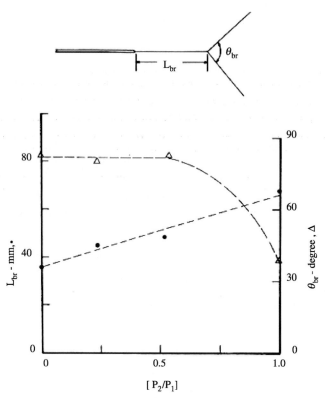

Figure 11.14 Variation of the branching angle with compressive crack parallel stress component. (Reproduced from Ravi-Chandar and Knauss, 1984d.)

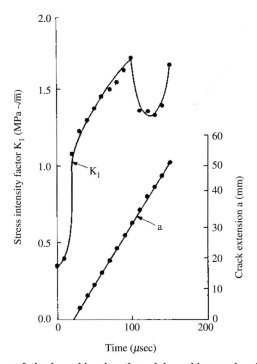

Time (μsec)

Figure 11.15 Variation of the branching length and branching angle with compressive crack parallel stress component. (Reproduced from Ravi-Chandar and Knauss, 1984d.)

tensile; the resulting branch angle θ_{br} increases by a large amount, with an included angle of about 135°.

The secondary stress pulse also causes a temporary decrease in the stress intensity factor, which is illustrated in Fig. 11.15 for the case when $P_1 = 10.35$ MPa and $P_2 = 5.5$ MPa. This delays the onset of branching. Furthermore, the compressive stress parallel to the crack axis would impede off-axis microcrack growth and decrease the size of the process zone. Indeed, examination of the fracture surface in Fig. 11.16 shows that

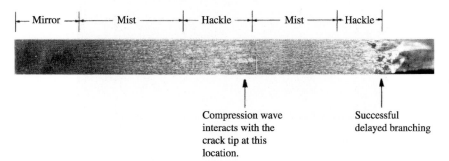

Figure 11.16 Influence of crack parallel compressive stress on fracture surface roughness evolution. (Reproduced from Ravi-Chandar and Knauss, 1984d.)

the surface roughness decreases upon arrival of the secondary stress pulse. Under the primary loading, the mirror, mist, hackle type fracture surface develops, but associated with the small drop in the stress intensity factor with the arrival of the secondary stress wave, the surface roughness decreases into the mist zone. Eventually, the loading from the primary pulse increases the stress intensity factor and results in branching; associated with this, the fracture surface roughness increases just prior to branching.

At this point, the phenomenology of branching seems quite clear: when a crack reaches a critical stage identified macroscopically by its stress intensity factor, it splits into two or more branches, each propagating with the same speed as the parent crack, but with a much reduced process zone. This is a clear indication that the process of branching is governed by the inner problem and not the outer problem that is treated by the continuum elastodynamics. In order to reveal the mechanisms underlying crack branching, Ravi-Chandar and Knauss (1984b) performed another experiment in which a high-speed camera was trained on a 3 mm diameter field of view located at the anticipated crack branching location; since the electromagnetic loading scheme was extremely repeatable, it was possible to estimate this location with reasonable accuracy. Fig. 11.17 shows a high-speed photomicrograph of the branching process—captured in the act of branching! Many

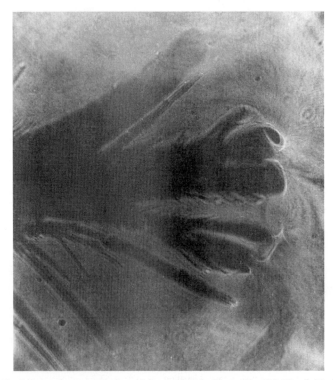

Figure 11.17 Real-time micrograph of crack branching in Homalite-100. The field of view is 3 mm. Note the large number of microbranches that are generated; three successful macrobranches emanated from this region. (Reproduced from Ravi-Chandar and Knauss, 1984b.)

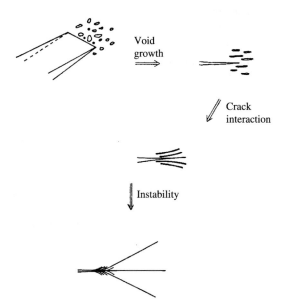

Void growth

Crack interaction

Instability

Figure 11.18 Mechanism of crack branching. (Reproduced from Ravi-Chandar and Knauss, 1984c.)

attempted branches are observed in this micrograph; these emanate from the main crack plane and continue to turn more or less smoothly away from the main direction of the crack. The micrographs indicate that (i) the branching process starts from microcracks that are initially parallel to the main crack propagation direction; (ii) these microcracks are not located preferentially through the plate thickness, thus giving branching a three-dimensional character; and (iii) the growth of successful microbranches is along a path smoothly turning away from the direction of the main crack.

This micrograph together with the real-time micrographs of the crack fronts reveal that the mechanism of crack growth and branching is multiple-microcracking. Ravi-Chandar and Knauss (1984c) used these observations to propose a mechanism for crack branching illustrated in Fig. 11.18. Initially, a crack propagates at the level of the initiation stress intensity factor generating a mirror-like fracture surface. The crack cuts through voids that may be present or nucleated by the crack tip stress field, with some of the voids diverting the crack to propagate along different planes; these are the origins of surface roughening. When the stress intensity level becomes sufficiently high, the voids grow into microcracks well ahead of the arrival of the main crack; this interaction leads to the well-known conic markings on the fracture surface. The idea of a single crack front is no longer applicable at the scale of the fracture process zone. The microcracks within the fracture process zone interact with each other and under suitable conditions repel each other; these deviated microcracks then appear as microbranches; these microbranches are observed in the real-time micrograph shown in Fig. 11.18 and have been studied in great detail recently by Fineberg and Marder (1999) and Hauch and Marder (1998).

Chapter 12

Phenomenological Models of Dynamic Fracture

Physical aspects of the generation of crack surface roughening considered in the previous section required postulating different models or mechanisms of crack growth. These mechanistic models have been developed further in the literature through models of the fracture phenomena; while some of these are derived from the micromechanics, others are approximate representations of the fracture process zones that do not require any mechanistic motivation. Such models range from discrete models of molecular dynamics and lattices to cohesive zone models of the fracture process to continuum damage models, thus covering scales from the atomic to continuum. While these models have been shown to be quite powerful, none of these models has been developed to the extent of the continuum theory. Specifically, these models are not yet capable of fully capturing the essence of the physical aspects of fracture discusses in the previous chapters. In this section we describe briefly some of the phenomenological models.

12.1 Discrete Models—Molecular Dynamics and Lattice Models

Discrete models of fracture have been considered at many length scales. At the smallest length scale, fracture—dynamic or static—results from the breakage of bonds between atoms. Thus, the hope is that simulations of cracking in a regular lattice arrangement of atoms with a known interaction potential should provide an indication of how fast fracture evolves. Thus, in *molecular dynamics* (MD) simulations, typically, a large number of atoms (about a million or two) are arranged usually in a perfect two-dimensional crystalline arrangement, and the motion and interaction of the atoms are calculated numerically using the assumed interaction potential. Abraham et al. (1994, 1997, 2003) used an idealized Lennard-Jones solid; Nakano et al. (1995) considered a porous silica and a silicon nitride in their simulations and included more complex interactions between the atoms. Since the numerical computation is time-consuming and expensive, the simulations are usually performed only over the size scale of a few nanometers spatially and only over a few picoseconds temporally. These computations are not yet extendable to realistic length and time scales; proper scaling of the results of the simulations requires mesoscopic models of the fracture process zone that are not yet available. Linking of the MD simulations directly to macroscale models through finite element methods is not an appropriate way to scale the results of the MD simulations since mesoscale structure and its influence on the fracture process evolution are then ignored; while this may be acceptable in determining global structure-independent properties, it is not likely to work for structure and scale-dependent phenomenon, like fracture. The results of MD simulations exhibit many of the features observed in experiments:

  - the average crack tip speed reaches a limiting value of about $0.57C_R$,
  - the instantaneous crack tip exhibits erratic oscillations due to crack path deflections beyond a crack speed of about $0.32C_R$, and
  - the crack surface exhibits significant roughness, caused initially by crack deflection from the tips and later by secondary cracks forming away from the main crack at an angle and linking with the main crack.

While these models appear to exhibit some similarities to experimentally observed behavior, these results are, however, somewhat paradoxical. The simulation is based on a regular arrangement of the atoms and hence should be predictive of cleavage fracture along crystallographic planes; but experiments indicate that the crack surface in single crystals is smooth and the crack speed reaches a significant fraction of the Rayleigh wave speed without exhibiting branching; of course, anisotropy of the fracture energy plays a crucial role in crystalline materials and could be a significant factor. On the other hand, in noncrystalline materials fast fracture exhibits crack surface roughening, limiting crack speed of about 50% of the Rayleigh wave speed and crack branching, all of which are present in the atomistic simulation. Thus, the simulation presents features that are observed at a much larger scale in experiments with noncrystalline materials! An issue still to be resolved is the role of discretization on the results; the discretized equations have a dynamics of their own, and their relationship to fast fracture in amorphous materials needs to be examined carefully. For instance, is the crack path instability simply a manifestation of the Yoffe (1951) stress field rearrangement in the discretized problem?

Lattice dynamics models of fast crack propagation have been presented by Slepyan and Fishkov (1981) and Slepyan (2002); Marder and Gross (1995) examined this model further to evaluate steady-state solutions and their stability. In these models, growth of a crack in an idealized spring mass lattice is considered, with a prescribed model for the spring behavior. We illustrate lattice dynamics with a mode III problem in a square lattice shown in Fig. 12.1.

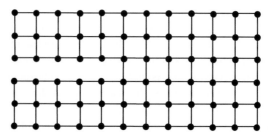

Figure 12.1 Square lattice with a crack. The lattice spacing is *a*.

The governing equation for the lattice can be obtained as a finite difference approximation of the anti-plane shear problem discussed in Eq. 3.8. Thus,

$$\frac{u_{m+1,n} - 2u_{m,n} + u_{m-1,n}}{a^2} + \frac{u_{m,n+1} - 2u_{m,n} + u_{m,n-1}}{a^2} = \frac{\rho}{\mu} \frac{d^2 u_{m,n}}{dt^2} \tag{12.1}$$

If the mass is considered to be concentrated at the nodes, the equivalent spring-mass model is obtained by setting $\rho a^2 = M$; thus the square lattice is governed the equation:

$$\mu(u_{m+1,n} + u_{m-1,n} + u_{m,n+1} + u_{m,n-1} - 4u_{m,n}) = M \frac{d^2 u_{m,n}}{dt^2} \tag{12.2}$$

This equation is valid for all rows except for the two rows immediately on the crack. Marder and Gross (1995) added a damping term and provided a fracture criterion for the lattice—that the lattice will break when the displacement reached a failure value, $2u_f$. These equations are then transformed into Fourier space and solved through a Wiener-Hopf procedure; see Slepyan (2002) and Marder and Gross (1995) for details. Marder and Gross used this model to examine the existence and stability of steady-state solutions. Their main result is shown in Fig. 12.2 where the crack speed normalized by the lattice speed is plotted as a function of the applied load, Δ. The main conclusions from the lattice solution are:
- There exist linearly stable lattice-trapped states corresponding to no crack growth, even though the load has exceeded the strength of the lattice. Such states were first discussed by Thompson (1971).
- Steady-state crack growth was not possible in the lattice at slow speeds; Marder and Gross called this the 'velocity gap' or the range of 'forbidden velocities'.
- Steady-state crack growth was possible in a range of speeds from about 0.3 to about 0.7 of the lattice wave speed.
- Steady-state crack growth was path unstable at very high speeds; Marder and Gross interpreted this as an indication of a limiting speed, since the path instabilities will lead eventually to crack branching.

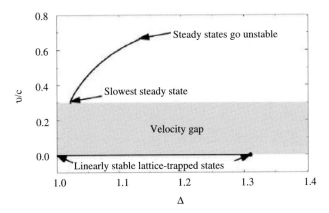

Figure 12.2 Crack velocity vs loading parameter Δ in a lattice. The lines indicate possible steady-state solutions. (From Fineberg and Marder, 1999.)

Lattice dynamics solutions present an interesting array of possibilities in modeling dynamic fracture; however, appropriate interpretation of the implications of these predictions to the nominally brittle amorphous materials that have been discussed here requires additional physical understanding and analytical modeling efforts. For example, how does the existence of a range of forbidden velocities in the lattice correlate with experimental observations? Dynamic fracture experiments indicate that in crack arrest experiments the crack speed can decrease smoothly from a significant fraction of the Rayleigh wave speed to zero (Dally, 1979; Kalthoff et al., 1980a,b), apparently going through the range of unstable speeds. Lattice trapping has also not been observed or demonstrated. Also, the lattice dynamics model reproduces many of the features of the MD simulations, and the paradox discussed earlier remains.

12.2 Cohesive Zone Models

Barenblatt (1959) posed the idea of a crack tip cohesive zone to account for the 'inner' problem of the fracture process; a similar model was suggested by Dugdale (1960) to model the line plastic zone at a crack tip. Since then generalizations of the cohesive zone ideas to craze failure in polymers (Knauss, 1974; Schapery, 1975) and general cohesive failure in weakening solids such as concrete by Hillerborg et al. (1976) have been suggested. While the physical motivations for postulating a cohesive model might be quite different in these different applications—ductile void growth and associated softening in metallic materials to microcracking in brittle materials such as concrete—the form of the cohesive model is similar in all cases: separation of the cohesive surfaces is described by imposing a constitutive relation between the traction vector connecting the cohesive surfaces and the displacement jump across the surfaces.

Xu and Needleman (1994) adapted the cohesive zone model for simulations of dynamic crack growth problem in brittle solids and incorporated it into a finite element formulation. They considered the cohesive model to be given in terms of a potential $\phi(\mathbf{\Delta})$, where $\mathbf{\Delta} = w_n \mathbf{n} + w_t \mathbf{t}$ is the cohesive surface separation; $\mathbf{n}$ and $\mathbf{t}$ the unit vectors in the normal and tangential directions, respectively; and w_n and w_t the normal and tangential cohesive surface separations, respectively. The traction-separation relation is then expressed as $\mathbf{T} = \partial\phi/\partial\mathbf{\Delta}$, where $\mathbf{T} = T_n \mathbf{n} + T_t \mathbf{t}$ and T_n, T_t are the normal and tangential components of the traction vector, respectively. The advantage of this formulation is that the energy balance in Eq. 3.6 may now be written incorporating the energy of the cohesive surfaces

$$\int_{\partial R} \mathbf{s} \cdot \frac{\partial \mathbf{u}}{\partial t} \, dR = \frac{\partial}{\partial t} \int_R [U(t) + T(t)] dV + \frac{\partial}{\partial t} \int_S \phi \, dS \tag{12.3}$$

where S is the area of the cohesive surfaces. The power supplied by the external tractions is then contained in the kinetic, potential and cohesive energies. Also, the cohesive surface model may be incorporated easily into the virtual work equation and used in formulating the discretized equations. In order to complete the model, the particular form of $\phi(\mathbf{\Delta})$ must be defined; Xu and Needleman (1994) assumed a form that resulted in a traction-separation relation that mimics interatomic separation processes. Their model is defined by four constants: σ_{max} and τ_{max}, the normal and tangential strengths of the cohesive

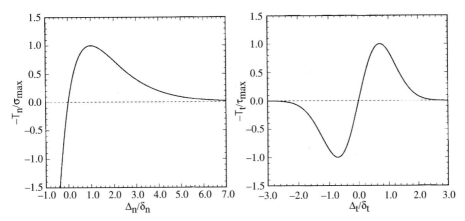

Figure 12.3 Cohesive surface traction-separation relationship; $\sigma_{max} = E/10 = 324$ **MPa,** $\tau_{max} = 755.5$ **MPa,** $\phi_n = \phi_t = 352.3$ **J/m^2. (From Xu and Needleman, 1994.)**

surface, respectively, and ϕ_n and ϕ_t, the work of separation under normal and shear tractions, respectively. The forms of the traction-separation law for crack opening and crack sliding are shown in Fig. 12.3. It should be noted that the form of the traction-separation relation derived from a potential of the type indicated above is conservative and thus this cohesive model is reversible; however, when it is used in simulations of crack growth, unloading occurs as the crack grows and hence healing of cohesive surfaces occur in this model. Irreversible models of the cohesive zone have been proposed, all based on the maximum attained crack surface separation (Geubelle and Rice, 1995; Yang and Ravi-Chandar, 1996; Ortiz and Pandolfi, 1999). For example, Yang and Ravi-Chandar used the following form

$$\mathbf{T} = k(w_d)\mathbf{\Delta} \tag{12.4}$$

where w_d is the maximum separation distance between two originally coincident points on the crack over the entire loading history, and is used as a damage parameter. The stiffness of the cohesive zone material $k(w_d)$ is assumed to depend on the current state of damage.

In their numerical simulations, Xu and Needleman (1994) discretized the elastic body into triangular elements, arranged in square or quadrilateral patterns. In order to model the cracking process, all element boundaries were connected to each other by cohesive surface elements. As a result of prescribing the cohesive surfaces, crack separation appears naturally in this simulation, just as in the case of the atomic and lattice models. However, as a result of the assumed form of $\phi(\mathbf{\Delta})$, no clear crack tip exits. The locations where a specific value of w_n is attained are identified as the crack tips; the time variation of the crack tips are tracked to determine the crack speed.

Crack speeds determined from one set of Xu and Needleman's simulations are shown in Fig. 12.4. The cohesive zone parameters used in these simulations are the following: $\sigma_{max} = E/10 = 324$ MPa, $\tau_{max} = 755.5$ MPa, $\phi_n = \phi_t = 352.3$ J/m^2. The elastic properties of the material were taken to correspond to that of PMMA: $E = 3.24$ GPa, $\nu = 0.35$, $\rho = 1190$ Mg/m^3, $C_d = 2090$ m/s, $C_s = 1004$ m/s and $C_R = 938$ m/s. Since cohesive

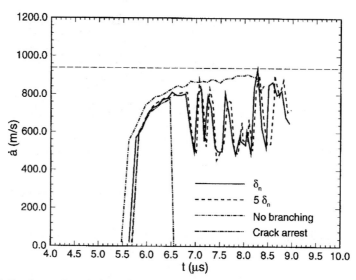

**Figure 12.4 Crack speed variation with time obtained in different simulations. 'No branching'
indicates simulation where cohesive surfaces were placed only along the prospective crack plane; the
resulting crack accelerates to the Rayleigh wave speed. 'Crack arrest' identifies simulation in which
the cohesive strength was increased abruptly resulting in an abrupt arrest of the crack. The other
two lines indicate simulations in which the cohesive surfaces were introduced at all element
boundaries. (From Xu and Needleman, 1994.)**

surfaces were provided in all element boundaries, arbitrary crack path evolution could be
obtained; crack path prediction from one simulation that indicated crack branching is
shown in Fig. 12.5. The main results from the numerical simulations are:

- The crack accelerates quickly to the Rayleigh wave speed if the cohesive surfaces
  are restricted to the initial crack plane; this is shown by the line identified by the
  label 'no branching' in Fig. 12.4.
- The instantaneous crack tip speed exhibits erratic oscillations due to crack path
  deflections beyond a crack speed of about $0.45C_R$; cohesive surfaces off the main
  crack plane were observed to separate and close as the main crack propagated. The
  oscillations appear to be related to the off-axis cohesive surface development, similar
  to the microbranching-induced oscillations observed by Fineberg et al. (1992).
- Very large stress levels are found to occur over some region near the crack; as a
  result, cohesive surfaces not connected with the main crack were found to exhibit
  opening displacement jumps reminiscent of microcrack nucleation and growth.
- Successful crack branching appears in the simulations.

The results of these simulations are also remarkably similar to the results from the MD
simulations. However, this is not surprising if one examines the underlying similarities in
the models. In both cases, one has a discrete set of (nodal or lattice) points, each with two
displacement degrees of freedom, u_α. The evolution equation for these displacement
degrees of freedom is described by the simple equation: $M\ddot{u}_\alpha = F(u_\alpha)$, where M is the mass
associated with the nodal or lattice point and F the force of interaction that depends on the

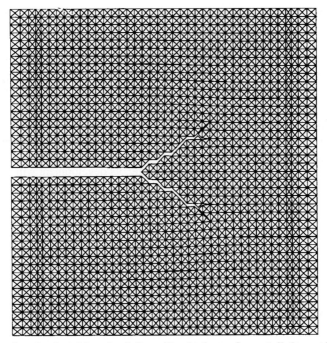

Figure 12.5 Crack path selected in a simulation with cohesive surfaces at all element boundaries and with symmetric loading; crack branching appears automatically. (From Xu and Needleman, 1994.)

displacement. The difference arises in the computation of the force: in the MD simulations, the force arises from the interaction with lattice points in a certain neighborhood and is calculated using the assumed interaction potential. In the finite-element simulations, the force arises from the elastic interaction with the surrounding nodes, as determined from the usual elastic analysis and from the cohesive law. From the similarity of the governing equation, one expects a similarity in the result from the two models.

Xu and Needleman (1994) also examined the influence of the discretization on the development of fast fracture. The results of this examination are perhaps the most useful in interpreting the results of all the discrete models discussed in this section. Xu and Needleman performed a number of simulations for a center crack in a strip of width 10 mm and height 2 mm; a velocity boundary condition of ± 10 m/s on the top and bottom boundaries was prescribed, with material properties appropriate for PMMA as described above. The resulting crack growth was examined in simulations with meshes where the cohesive boundaries off the initial crack plane were oriented at angles of $\pm 15°$, $\pm 30°$, $\pm 45°$ and $\pm 60°$. They made the following observations from the results of the simulations:

– when the cohesive surfaces are at $\pm 15°$ and $\pm 30°$, the crack grows in a zigzag mode with crack speed oscillations appearing right from crack initiation;
– crack speed oscillations appear to be most pronounced when the cohesive surfaces are at $\pm 45°$;

- crack speed oscillations are almost completely absent when the cohesive surfaces are at $\pm 60°$;
- onset of crack branching is affected significantly by the angle of the cohesive surfaces; the crack speed at branching varies nonmonotonically with the angle of the cohesive surfaces;
- in almost all the cases, the crack speed continues to increase and reaches values very close to the Rayleigh wave speed; for cohesive surfaces at $\pm 60°$, at the onset of branching, the calculated crack speed is $0.91 C_R$.

These observations are consistent with the Yoffe type instability of the crack path; at high crack speeds, the maximum tensile stresses arise off the initial crack plane. If the cohesive surface is oriented at the appropriate angle, its development will be enhanced as the crack speed increases; the crack branches and velocity oscillations occur. If, on the other hand, the cohesive surface orientation is not at the appropriate angle, its development will be suppressed and the crack continues to grow straight ahead at high speeds.

An important point to note is that crack extension occurs only along the element boundaries; the underlying assumption in performing such simulations is that by making the element size small, the simulation can model arbitrary evolution of the crack path. In a recent evaluation of this strategy Falk et al. (2001) have questioned this proposition by pointing out that, in simulations aimed at analyzing crack branching, the mesh size introduced a characteristic length scale into the problem that influenced the results.

12.3 Continuum Damage Models

Macroscopic or continuum level modeling of the fracture process zone has also been attempted. Johnson (1992) considered a continuum damage mechanics approach following a cell model for fracture developed by Broberg (1982); in the region near the crack tip the material was considered to be of an elastic-softening type and the damage was assumed to be triggered by the dilatational strain. Thus, in the crack tip region, the elastic stiffness of each element was decreased by a factor that depended on the current level of dilatation, ϑ

$$
\omega(\nu) = \begin{cases} 1 & 0 < \vartheta < \vartheta_1 \\ \dfrac{(\vartheta_2/\vartheta)^n - 1}{(\vartheta_2/\vartheta_1)^n - 1} & \vartheta_1 < \vartheta < \vartheta_2 \\ 0 & \vartheta > \vartheta_2 \end{cases} \tag{12.5}
$$

where ϑ_1 is a threshold dilatation below which no damage occurs, ϑ_2 the dilatation at which the damage is complete and n a parameter. Irreversibility of damage was also incorporated into the model; if the volume dilatation decreased temporarily after exceeding the damage threshold, ω was maintained at its previous maximum value. This model was implemented in a very fine-scale finite-element simulation of the pressurized semi-infinite crack problem using the same conditions that were attained in the experiments of Ravi-Chandar and Knauss described in Section 9.3. While the scale of the elements is large compared to the atomic scale fracture processes, the mesh size was

large enough to capture the structure of the overall fracture process zone. Because of computational cost considerations, Johnson introduced the damage model only in the 12 rows immediately next to the crack surface; this restriction results in some artifacts in the numerical simulations, but these are discriminated easily. The main results from this numerical simulation of the continuum damage model are:
  – In each simulation, the crack grew at a constant speed that depended on the applied load; concomitant with this constant speed was an appropriate expansion of the process zone, identified as cells in which the dilatation has exceeded either ϑ_1 or ϑ_2; qualitative correspondence between the simulations and experiments described in connection with Figs. 9.2, 9.3 and 9.4 was demonstrated. In Fig. 12.6 the evolution of the damage in the cells as a result of crack growth is shown. At a crack surface

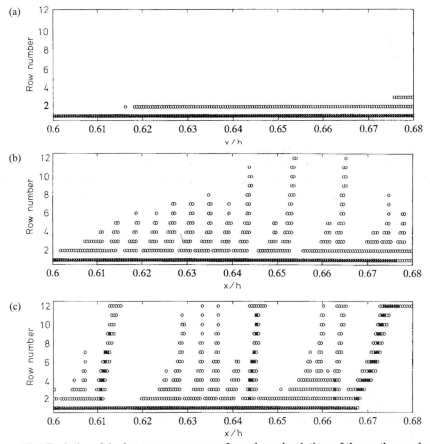

Figure 12.6 Evolution of the fracture process zone from three simulations of the continuum damage model. $\vartheta_1 = 0.00075$, $\vartheta_2 = 0.018$ and $n = 1.5$; the crack surface pressure was $0.001E$ in (a), $0.002E$ in (b) and $0.003E$ in (c). The circles indicate elements that have begun to accumulate damage (max $\vartheta > \vartheta_1$) and the crosses indicate elements in which damage is complete (max $\vartheta > \vartheta_2$). (Reproduced from Johnson, 1992.)

pressure of $0.001E$, the process zone developed only along the first three rows of elements. As the crack surface pressure was increased to $0.003E$, the process zone began to develop faster in the direction normal to the crack line and left a trail of partially or fully damaged material.

- With increased crack surface loading, the crack accelerated to a limiting speed that is significantly smaller than the Rayleigh wave speed.
- If the spread of the damaged elements near the crack tip is identified as crack surface roughening, increase in surface roughness was correlated with energy supply rather than velocity.
- If the development of damage concentrated along specific directions other than the initial crack line is considered as indicative of branching, this simulation also showed the development of branching; however, whether this is driven by a Yoffe-type instability or not is not clear. If should be recognized that the local properties of the damaged cells vary significantly from the initial properties and hence the crack speed might be large enough to trigger the Yoffe-type instability. Recent work by Klein and Gao (2001) has suggested that this might be the case.

The results exhibit many of the characteristics observed in experiments. This type of a model is quite attractive since it is not computationally as intensive as the discrete models such as MD simulations and furthermore, the simulation could be performed easily in standard finite element codes. However, additional work is necessary to determine the influence of discretization on the results and the appropriate form of the damage law to be used. For instance, the damage law could be derived from homogenized description of the microcracked crack tip material. Such damage models based on nucleation and growth models for crack growth were considered first by Zhurkov (1965) for thermally activated crack processes such as creep. Curran et al. (1973) applied these ideas to the stress-induced nucleation and growth processes in the dynamic spalling problem. The description of dynamic crack growth provided by Ravi-Chandar and Knauss (1984a–d) and Ravi-Chandar and Yang (1997) is also based on nucleation and growth of cavities and microcracks. In these models, the nucleation rate of microcracks is given by

$$\dot{N} = \dot{N}_0 \exp[(\sigma - \sigma_{n0})/\sigma_1] \qquad (12.6)$$

where σ_{n0} is the nucleation threshold, and σ_1 and $\dot{N}_0$ the constants to be determined through calibration experiments. Note that this form of the nucleation rate is typically applied in many models, not necessarily related to fracture. Zhurkov (1965) applied this to the rate of bond breakage and Curran et al. (1973) applied it to nucleation of microcracks. Based on experimental observations, Curran et al. also assumed an exponential distribution of initial radii of the nucleated microcracks: $\Delta N(R) = \Delta N_0 \exp[-R/R_1]$, where $\Delta N(R)$ is the number of microcracks per unit volume with radius larger than R that are nucleated in a time interval Δt and R_1 is another parameter in the model. Finally, a growth rate is imposed on the microcracks such that

$$\frac{\dot{R}}{R} = \left[\frac{\sigma - \sigma_0(R)}{4\eta} \right] \qquad (12.7)$$

where $\sigma_0(R)$ is the critical stress for a penny-shaped flaw of radius R to be fracture critical and η the viscosity; this form was derived by Curran et al. from viscosity-limited growth in ductile materials, but was also applied successfully to brittle materials. With this nucleation and growth model, for any load history, the current state of damage can be determined and used in degrading the material properties for simulating the overall response of the material. Such models of the mechanical properties of the damaging material could then be incorporated into numerical simulations of the fracture problem in the sense of the continuum simulations of Johnson (1992) or the cohesive zone simulations of Xu and Needleman (1994) and others.

In this chapter, we have reviewed some of the phenomenological models that are aimed at providing an analysis of the "inner problem" associated with the dynamics of the fracture process zone. While there have been significant advances in the computational methods used, comparison to experimental results remain qualitative at best. Fundamental considerations regarding the determination of appropriate material properties for inclusion in these phenomenological models and the quantitative comparison of the results of the simulations to experimental observations remain open issues.

References

F.F. Abraham, D. Brodbeck, R.A. Rafey and W.F. Rudge: Instability Dynamics of Fracture: A Computer Simulation Investigation, Phys. Rev. Lett., **73** (1994), 272.

F.F. Abraham, D. Brodbeck, W.E. Rudge and X. Xu: A Molecular Dynamics Investigation of Rapid Fracture Mechanics, J. Mech. Phys. Solids, **45** (1997), 1595–1619.

F.F. Abraham: How Fast Can Cracks Move? A Research Adventure in Materials Failure Using Millions of Atoms and Big Computers, Adv. Phys., **52** (2003), 727–790.

J.D. Achenbach: Wave Propagation in Elastic Solids, North-Holland Publishing Company, Amsterdam, 1973.

S.R. Anthony, J.P. Chubb and J. Congleton: The Crack Branching Velocity, Philos. Mag., **22** (1970), 1201–1261.

K. Arakawa and K. Takahashi: Relationship Between Fracture Parameters and Surface Roughness of Brittle Polymers, Int. J. Fract., **48** (1991), 103–114.

A.S. Argon and M.M. Salama: Growth of Crazes in Glassy Polymers, Phil. Mag., **36** (1977), 1217–1234.

A. Assa, J. Politch and A.A. Betser: Slope and Curvature Measurement by a Double-Frequency-Grating Shearing Interferometer, Exp. Mech., **19** (1979), 129–137.

ASTM E1221-96: Standard Test Method for Determining Plane-Strain Crack-Arrest Fracture Toughness, K_{Ia}, of Ferritic Steels.

C. Atkinson and J.D. Eshelby: The Flow of Energy into the Tip of a Moving Crack, Int. J. Fract., **4** (1968), 3–8.

B.R. Baker: Dynamic Stresses Created by a Moving Crack, J. Appl. Mech., **29** (1962), 449–458.

G.I. Barenblatt: Concerning Equilibrium Cracks Forming During Brittle Fracture: The Stability of Isolated Cracks, Relationship with Energetic Theories, Appl. Math. Mech., **23** (1959), 1273–1282 (English Translation of PMM, **23** (1959), 893–900).

B.R. Bass, C.E. Pugh and H.K. Stamm: Dynamic Analysis of a Crack Run-Arrest Experiment in a Nonisothermal Plate, Pressure Vessel Components Design and Analysis, ASME PVP98-2, 1985, pp. 175–184.

W.M. Beebe: An Experimental Investigation of Dynamic Crack Propagation in Plastic and Metals, Ph.D Thesis, California Institute of Technology, Pasadena, CA, 1966.

J.R. Berger, J.W. Dally and R.J. Sanford: Determining the Dynamic Stress Intensity Factor with Strain Gages Using a Crack Tip Locating Algorithm, Engng Fract. Mech., **36** (1990), 145–156.

H. Bergkvist: Some Experiments on Crack Motion and Arrest in Polymethylmethacrylate, Engng Fract. Mech., **6** (1974), 621–626.

W. Böhme: Dynamic Key-Curves for Brittle Fracture Impact Tests and Establishment of a Transition Time, In: J.P. Gudas, J.A. Joyce and E.M. Hackett (Eds): Fracture Mechanics: Twenty-First Symposium, ASTM STP 1074, American Society for Testing and Materials, Philadelphia, 1990, pp. 144–156.

D. Bonamy and K. Ravi-Chandar: Interaction of Stress Waves with Propagating Cracks, Phys. Rev. Lett, **91** (2003), 235502.

M. Born and E. Wolf: Principles of Optics, 7th expanded ed., Cambridge University Press, Cambridge, 1999.

E. Bouchaud: Scaling Properties of Cracks, J. Phys. Condens. Matter, **9** (1997), 4319–4344.

F.P. Bowden, J.H. Brunton, J.E. Field and A.D. Hayes: Controlled Fracture of Brittle Solids and Interruption of Electric Current, Nature, **216** (1967), 38–42.

W.B. Bradley and A.S. Kobayashi: An Investigation of Propagating Cracks by Dynamic Photoelasticity, Exp. Mech., **10** (1970), 106–113.

D. Brewster: On the Affections of Light Transmitted Through Crystallized Bodies, Philos. Trans., **104** (1814), 187–218.

D. Brewster: On the Effects of Simple Pressure in Producing that Species of Crystallization Which Forms Two Oppositely Polarized Images, and Exhibits the Complementary Colors by Polarized Light, Philos. Trans., **105** (1815), 60–64.

B. Brickstad: A FEM Analysis of Crack Arrest Experiments, Int. J. Fract., **21** (1983), 177–194.

B. Brickstad and F. Nilsson: Dynamic Analysis of Crack Growth and Arrest in a Pressure Vessel Subjected to Thermal and Pressure Loading, Engng Fract. Mech., **23** (1986a), 61–70.

B. Brickstad and F. Nilsson: A Dynamic Analysis of Crack Propagation and Arrest in Pressurized Thermal Shock Experiments, Engng Fract. Mech., **23** (1986b), 99–102.

K.B. Broberg: The Propagation of a Brittle Crack, Ark. Fysik, **18** (1960), 159–192.

K.B. Broberg: Foundations of Fracture Mechanics, Engng Fract. Mech., **16** (1982), 497–515.

K.B. Broberg: The Near-Tip Field at High Crack Velocities, Int. J. Fract., **39** (1989), 1–13.

K.B. Broberg: Cracks and Fracture, Academic Press, New York, 1999.

H.D. Bui, H. Maigre and D. Rittel: A New Approach to the Experimental Determination of the Dynamic Stress Intensity Factor, Int. J. Solids Struct., **29** (1992), 2881–2895.

R. Burridge: An Influence Function for the Intensity Factor in Tensile Fracture, Int. J. Engng Sci., **14** (1976), 725–734.

R. Burridge, G. Conn and L.B. Freund: The Stability of a Rapid Mode II Shear Crack with Finite Cohesive Traction, J. Geophys. Res., **84** (1979), 2210–2222.

J. Carlsson, L. Dahlberg and F. Nilsson: Experimental Studies of the Unstable Phase of Crack Propagation in Metals and Polymers, In: G.C. Sih (Ed): Dynamic Crack Propagation, Noordhoff International Publishing, Leyden, 1973, pp. 165–181.

Y.J. Chao, M.A. Sutton and C.E. Taylor: Interferometric Determination of Curvatures of Flexed Plate, J. Appl. Mech., **49** (1982), 837–842.

M.G. Charpy: Note sur L'essai des Métaux, Mémoires et Compte-rendus de la Société des Ingénieurs Civils de France, June 1901, pp. 848–877. English translation reprinted in T.A. Siewert and M.P. Nanahan, Sr. (Eds.): Pendulum Impact Testing: A Century of Progress, ASTM STP 1380, American Society for Testing and Materials, West Conshohocken, PA, 2000, pp. 46–69.

E.P. Chen and G.C. Sih: Transient Response of Cracks to Impact Load, Mechanics of Fracture, Volume IV: Elastodynamic Crack Problems, Noordhoff International Publishing, Leyden, 1977, pp. 1–58.

R.D. Cheverton, S.E. Bolt, P.P. Holz and S.K. Iskander: Behavior of Surface Flaws in Reactor Pressure Vessels Under Thermal-Shock Loading Conditions, Exp. Mech., **21** (1981), 155–162.

J. Congleton and N.J. Petch: Crack-Branching, Philos. Mag., **16** (1967), 749–760.

L.S. Costin, J. Duffy and L.B. Freund: Fracture Initiation in Metals Under Stress Wave Loading Conditions, In: G.T. Hahn and M.F. Kanninen (Eds): Fast Fracture and Crack Arrest, ASTM STP 627, American Society for Testing and Materials, Philadelphia, 1977, pp. 301–318.

B. Cotterell: Velocity Effects in Fracture Propagation, Appl. Mater. Res., **4** (1965), 227–232.

B. Cotterell: Fracture Propagation in Organic Glasses, Int. J. Fract. Mech., **4** (1968), 209–217.

B. Cotterell and J.R. Rice: Slightly Curved or Kinked Cracks, Int. J. Fract., **16** (1980), 155–169.

J.W. Craggs: On the Propagation of a Crack in an Elastic-Brittle Material, J. Mech. Phys. Solids, **8** (1960), 66–75.

T. Cramer, A. Wanner and P. Gumbsch: Crack Velocities During Dynamic Fracture of Glass and Single Crystalline Silicon, Phys. Stat. Solidi A, **164** (1997), R5–R6.

P.B. Crosley and E.J. Ripling: Dynamic Fracture Toughness of A533 Steel, J. Basic Engng, **91** (1969), 525–534.

P.B. Crosley and E.J. Ripling: Significance of Crack Arrest Toughness K_{Ia} Testing, Crack Arrest Methodology and Application—ASTM STP 711, American Society for Testing and Materials, Philadelphia, 1980, pp. 321–337.

D.R. Curran, D.A. Shockey and L. Seaman: Dynamic Fracture Criteria for a Polycarbonate, J. Appl. Phys., **44** (1973), 4025.

L. Dahlberg, F. Nilsson and B. Brickstad: Influence of Specimen Geometry on Crack Propagation and Arrest Toughness, Crack Arrest Methodology and Application—ASTM STP 711, American Society for Testing and Materials, Philadelphia, 1980, pp. 89–108.

J.W. Dally: Dynamic Photoelastic Studies of Fracture, Exp. Mech., **19** (1979), 349–361.

J.W. Dally and D.B. Barker: Dynamic Measurements of Initiation Toughness at High Loading Rates, Exp. Mech., **28** (1988), 298–303.

J.W. Dally and J.R. Berger: The Role of the Electrical Resistance Strain Gage in Fracture Research, In: J.S. Epstein (Ed): Experimental Techniques in Fracture, VCH, New York, 1993, pp. 1–39.

J.W. Dally and T. Kobayashi: Crack Arrest in Duplex Specimens, Int. J. Solids Struct., **14** (1978), 121–129.

J.W. Dally and W.F. Riley: Experimental Stress Analysis, 2nd ed., McGraw-Hill, New York, 1978.

W. Döll: Transition from Slow to Fast Crack Propagation in PMMA, J. Mater. Sci., **11** (1976a), 2348.

W. Döll: Application of an Energy Balance and an Energy Method to Dynamic Crack Propagation, Int. J. Fract., **12** (1976b), 595–605.

M.J. Doyle: A Mechanism for Crack Branching in Polymethyl Methacrylate and the Origin of the Bands on the Surfaces of Fracture, J. Mater. Sci., **18** (1983), 687–702.

D.S. Dugdale: Yielding of Steel Sheets Containing Slits, J. Mech. Phys. Solids, **8** (1960), 100–104.

E.N. Dulaney and W.F. Brace: Velocity Behavior of a Growing Crack, J. Appl. Phys., **31** (1960), 2233–2236.

H.E. Edgerton and F.E. Bartow: J. Amer. Ceram. Soc., **41**(1941), 131.

J. Eftis and J.M. Kraft: A Comparison of the Initiation with the Rapid Propagation of a Crack in a Mild Steel Plate, J. Basic Engng, Trans. ASME, 1965, 257–263.

J.S. Epstein and M.S. Dadkhah: Moire Interferometry in Fracture Research, In: J.S. Epstein (Ed): Experimental Techniques in Fracture, VCH, New York, 1993, pp. 427–508.

J.D. Eshelby: Energy Relations and the Energy-Momentum Tensor in Continuum Mechanics, In: M.F. Kanninen et al. (Eds): Inelastic Behavior of Solids, McGraw-Hill, New York, 1970, pp. 77–115.

M.L. Falk, A. Needleman and J.R. Rice: A Critical Evaluation of Cohesive Zone Models of Dynamic Fracture, J. Phys. IV, **11** (2001), 43–50.

J.E. Field: Brittle Fracture: Its Study and Application, Contemp. Phys., **12** (1971), 1–31.

J.E. Field: High-Speed Photography, Contemp. Phys., **24** (1983), 439–459.

J. Fineberg and M. Marder: Instability in Dynamic Fracture, Phys. Rep., **313** (1999), 1–108.

J. Fineberg, S.P. Gross, M. Marder and H.L. Swinney: Instability in Dynamic Fracture, Phys. Rev. Lett., **67** (1991), 457–460.

J. Fineberg, S.P. Gross, M. Marder and H.L. Swinney: Instability in the Propagation of Fast Cracks, Phys. Rev., **B45** (1992), 5146–5154.

L.B. Freund: Energy Flux into the Tip of an Extending Crack in an Elastic Solid, J. Elasticity, **2** (1972a), 341–349.

L.B. Freund: Crack Propagation in an Elastic Solid Subjected to General Loading. I. Constant Rate of Extension, J. Mech. Phys. Solids, **20** (1972b), 129–140.

L.B. Freund: Crack Propagation in an Elastic Solid Subjected to General Loading. III. Stress Wave Loading, J. Mech. Phys. Solids, **21** (1973), 47–61.

L.B. Freund: Crack Propagation in an Elastic Solid Subjected to General Loading. IV. Obliquely Incident Stress Pulse, J. Mech. Phys. Solids, **22** (1974a), 137–146.

L.B. Freund: Stress Intensity Factor Due to Normal Impact Loading of the Faces of a Crack, Int. J. Engng Sci., **12** (1974b), 179–189.

L.B. Freund: The Mechanics of Dynamic Shear Crack Propagation, J. Geophys. Res., **84** (1979), 2199–2209.

L.B. Freund: Dynamic Fracture Mechanics, Cambridge University Press, Cambridge, 1990.

L.B. Freund and A.J. Rosakis: The Structure of the Near-Tip Field During Transient Elastodynamic Crack Growth, J. Mech. Phys. Solids, **40** (1992), 699–719.

P.H. Geubelle and J.R. Rice: A Spectral Method for Three-Dimensional Elastodynamic Fracture Problems, J. Mech. Phys. Solids, **43** (1995), 1791–1824.

J.J. Gilman, C. Knudsen and W.P. Walsh: Cleavage Cracks and Dislocations in LiF Crystals, J. Appl. Phys., **6** (1958), 601–607.

K.F. Graff: Wave Motion in Elastic Solids, Ohio State University Press, Ohio, 1975.

A.K. Green and P.L. Pratt: Measurement of the Dynamic Fracture Toughness of Polymethylmethacrylate by High-Speed Photography, Engng Fract. Mech., **6** (1974), 71.

J.A. Hauch and M. Marder: Energy Balance in Dynamic Fracture, Investigated by a Potential Drop Technique, Int. J. Fract., **90** (1998), 133–151.

A. Hillerborg, M. Modeer and P.E. Peterson: Analysis of Crack Formation and Crack Growth in Concrete by Means of Fracture Mechanics and Finite Elements, Cem. Concr. Res., **6** (1976), 773–782.

R.G. Hoagland, A.R. Rosenfield, P.C. Gehlen and G.T. Hahn: A Crack Arrest Measuring Procedure for K_{Im}, K_{ID}, and K_{Ia} Properties, In: G.T. Hahn and M.F. Kanninen (Eds): Fast Fracture and Crack Arrest, ASTM STP 627, American Society for Testing and Materials, Philadelphia, 1977, pp. 177–202.

J. Hopkinson: Original Papers, Cambridge University Press, Cambridge, 1901, pp. 310–320.

D. Hull: Influence of Stress Intensity and Crack Speed on Fracture Surface Topography: Mirror to Mist Transition, J. Mater. Sci., **31** (1997a), 1829–1841.

D. Hull: Influence of Stress Intensity and Crack Speed on Fracture Surface Topography: Mirror to Mist to Macroscopic Bifurcation, J. Mater. Sci., **31** (1997b), 4483–4492.

D.R. Ireland: Procedures and Problems Associated with Reliable Control of the Instrumented Impact Test, Instrumented Impact Testing, ASTM STP 536, 1974, pp. 3–29.

G.R. Irwin and J.A. Kies: Fracturing and Fracture Dynamics, Weld. J., NY Res. Suppl., **31** (1952), 95.

K.D. Ives, A.K. Shoemaker and R.F. McCartney: Pipe Deformation During a Running Shear Fracture in Line Pipe, J. Engng Mater. Technol., **96** (1974), 309–317.

E. Johnson: Process Region Changes for Rapidly Propagating Cracks, Int. J. Fract., **55** (1992), 47–63.

R.C. Jones: A New Calculus for the Treatment of Optical Systems, J. Opt. Soc. Am., **31**, (1941) (Part I) 488–493, (Part II) 493–499.

J. Jung and M.F. Kanninen: An Analysis of Dynamic Crack Propagation and Arrest in a Nuclear Pressure Vessel Under Thermal Shock Conditions, J. Press. Vessel Technol., **105** (1983), 111–116.

J.F. Kalthoff: On Some Current Problems in Experimental Fracture Dynamics, In: W.G. Knauss, K. Ravi-Chandar and A.J. Rosakis (Eds): The Proceedings of the NSF-ARO Workshop on Dynamic Fracture, 1983, pp. 11–35.

J.F. Kalthoff: Fracture Behavior Under High Rates of Loading, Engng Fract. Mech., **23** (1986), 289–298.

J.F. Kalthoff: Shadow Optical Method of Caustics, In: A.S. Kobayashi (Ed): Handbook of Experimental Mechanics, Prentice Hall, New York, 1987, pp. 430–498.

J.F. Kalthoff: Experimental Fracture Dynamics, In: J.R. Klepaczko (Ed): Crack Dynamics in Metallic Materials, Springer, Vienna, 1990a, pp. 69–254.

J.F. Kalthoff: Shadow Optical Analysis of Dynamic Shear Fracture, Opt. Engng, **27** (1990b), 835–840.

J.F. Kalthoff and D.A. Shockey: Instability of Crack Under Impulse Loads, J. Appl. Phys., **48** (1977), 986–993.

J.F. Kalthoff, J. Beinert and S. Winkler: Measurement of Dynamic Stress Intensity Factors for Fast Running and Arresting Cracks in Double Cantilever Beam Specimens, In: G.T. Hahn and M.F. Kanninen (Eds): Fast Fracture and Crack Arrest, ASTM STP 627, American Society for Testing and Materials, Philadelphia, 1977, pp. 161–176.

J.F. Kalthoff, J. Beinert and S. Winkler: Analysis of Fast Running and Arresting Cracks by the Shadow-Optical Method of Caustics, IUTAM Symposium on Optical Methods in the Mechanics of Solids, University of Poitiers, France, 1980a, pp. 497–508.

J.F. Kalthoff, J. Beinert, S. Winkler and W. Klemm: Experimental Analysis of Dynamic Effects in Different Crack Arrest Test Specimens, Crack Arrest Methodology and Application—ASTM STP 711, American Society for Testing and Materials, Philadelphia, 1980b, pp. 109–127.

M. Kavaturu, A. Shukla and A.J. Rosakis: Intersonic Crack Propagation Along Interfaces: Experimental Observations and Analysis, Exp. Mech., **38** (1998), 218–225.

F. Kerhkof: Wave Fractographic Investigation of Brittle Fracture Dynamics, In: G.C. Sih (Ed): Dynamic Crack Propagation, Noordhoff International Publishing, Leyden, 1973, pp. 3–35.

J.A. Kies, A.M. Sullivan and G.R. Irwin: Interpretation of Fracture Markings, J. Appl. Phys., **21** (1950), 716–720.

K.S. Kim: Dynamic Fracture Under Normal Impact Loading of the Crack Faces, J. Appl. Mech., **52** (1985a), 585–592.

K.S. Kim: A Stress Intensity Factor Tracer, J. Appl. Mech., **52** (1985b), 291–297.

V.K. Kinra and C.L. Bowers: Brittle Fracture of Plates in Tension. Stress Field Near the Crack, Int. J. Solids Struct., **17** (1981), 175.

P.A. Klein and H. Gao: Study of Crack Dynamics Using the Virtual Internal Bond Method, In: T.-J. Chuang and J.W. Rudnicki (Eds): Multiscale Deformation and Fracture in Materials and Structures, Kluwer Academic Publishers, Dordrecht, 2001, pp. 275–309.

J.R. Klepaczko: Discussion of a New Experimental Method in Measuring Fracture Toughness Initiation at High Loading Rates by Stress Waves, J. Engng Mater. Technol., **104** (1982), 29–35.

J.R. Klepaczko: Fracture Initiation Under Impact, Int. J. Impact Engng, **3** (1985), 191–210.

J.R. Klepaczko: Dynamic Crack Initiation, Some Experimental Methods and Modeling, In: J.R. Klepaczko (Ed): Crack Dynamics in Metallic Materials, Springer, Vienna, 1990, pp. 255–454.

W.G. Knauss: On the Steady Propagation of a Crack in a Viscoelastic Sheet: Experiments and Analysis, In: H.H. Kausch et al. (Eds): Deformation and Fracture of High Polymers, Plenum Press, New York, 1974, pp. 501–541.

A.S. Kobayashi and S. Mall: Dynamic Fracture Toughness of Homalite-100, Exp. Mech., **18** (1978), 11–18.

A.S. Kobayashi, B.G. Wade and W.B. Bradley: Fracture Dynamics of Homalite-100, In: H.H. Hausch et al. (Eds): Deformation and Fracture of High Polymers, Plenum Press, New York, 1973, pp. 487–500.

A.S. Kobayashi, A.F. Emery and S. Mall: Dynamic Finite Element and Dynamic Photoelastic Analyses of Crack Arrest in Homalite-100 Plates, In: G.T. Hahn and M.F. Kanninen (Eds): Fast Fracture and Crack Arrest, ASTM STP 627, American Society for Testing and Materials, Philadelphia, 1977, pp. 95–108.

A.S. Kobayashi, K. Seo, J.Y. Jou and Y. Urabe: A Dynamic Analysis of Modified Compact-Tension Specimens Using Homalite-100 and Polycarbonate Plates, Exp. Mech., **20** (1980), 73–79.

T. Kobayashi and J.W. Dally: Dynamic Photoelastic Determination of the Relation for 4340 Steel, Crack Arrest Methodology and Application—ASTM STP 711, American Society for Testing and Materials, Philadelphia, 1980, pp. 189–210.

M. Kosai and A.S. Kobayashi: Axial Crack Propagation and Arrest in Pressurized Fuselage, In: S.N. Atluri, S.G. Sampath and P. Tong (Eds): Structural Integrity of Aging Aircraft, Springer, Berlin, 1991, pp. 225–239.

M. Kosai, A. Shimamoto, C.T. Yu, A.S. Kobayashi and P.W. Tan: Axial Crack Propagation and Arrest in a Pressurized Cylinder: An Experimental–Numerical Analysis, Exp. Mech., **39** (1999), 256–264.

B.V. Kostrov: Unsteady Crack Propagation of Longitudinal Shear Cracks, Appl. Math. Mech., **30** (1966), 1241–1248, (English Translation from PMM, **30** (1966), 1042–1049).

B.V. Kostrov: On the Crack Propagation with Variable Velocity, Int. J. Fract., **11** (1975), 47–56.

B.V. Kostrov and L.V. Nikitin: Some General Problems of Mechanics of Brittle Fracture, Arch. Mech. Stosowanej, **22** (1970), 749–775.

S. Krishnaswamy and A.J. Rosakis: On the Extent of Dominance of Asymptotic Elastodynamic Crack Tip Fields: Part I: An Experimental Study Using Bifocal Caustics, J. Appl. Mech., **58** (1990), 87–94.

S. Krishnaswamy, H.V. Tippur and A.J. Rosakis: Measurement of Transient Crack Tip Deformation Fields Using the Method of Coherent Gradient Sensing, J. Mech. Phys. Solids, **40** (1992), 339–372.

M.K. Kuo: Stress Intensity Factors for a Semi-infinite Plane Crack Under a Pair of Point Forces, J. Elasticity, **30** (1993), 197–209.

M.K. Kuo and T.Y. Chen: The Wiener-Hopf Technique in Elastodynamic Crack Problems with Characteristic Lengths in Loading, Engng Fract. Mech., **42** (1992), 805–813.

H. Lee and S. Krishnaswamy: A Compact Polariscope/Shearing Interferometer for Mapping Stress Fields in Bimaterial Systems, Exp. Mech., **36** (1996), 404–411.

Y.J. Lee and L.B. Freund: Fracture Initiation Due to Asymmetric Impact Loading of an Edge Cracked Plate, J. Appl. Mech., **57** (1990), 104–111.

J. Leeuwerik: Kinematic Features of the Brittle Fracture Phenomenon, Rheol. Acta, **2** (1962), 10–16.

C. Liu and A.J. Rosakis: On the Higher Order Asymptotic Analysis of a Non-uniformly Propagating Dynamic Crack Along an Arbitrary Path, J. Elasticity, **35** (1994), 27–60.

C. Liu, A.J. Rosakis and L.B. Freund: Interpretation of Optical Caustics in the Presence of Non-uniform Crack Tip Motion Histories: A Study Based on a Higher Order Transient Crack Tip Expansion, Int. J. Solids Struct., **30** (1993), 875–897.

C. Liu, W.G. Knauss and A.J. Rosakis: Loading Rate and Dynamic Initiation Toughness in Brittle Solids, Int. J. Fract., **90** (1998), 103–118.

A.E.H. Love: A Treatise on the Mathematical Theory of Elasticity, 4th ed., Cambridge University Press, Cambridge, 1927.

C.C. Ma: Analysis of the Caustic Method for the Transient Stress Field of a Stationary Crack, J. Appl. Mech., **58** (1991), 591–593.

C.C. Ma and L.B. Freund: The Extent of the Stress Intensity Factor Field During Crack Growth Under Dynamic Loading Conditions, J. Appl. Mech., **53** (1986), 303–310.

R.V. Mahajan and K. Ravi-Chandar: Experimental Determination of Stress Intensity Factors Using Caustics and Photoelasticity, Exp. Mech., **29** (1989), 6–11.

H. Maigre and D. Rittel: Dynamic Fracture Detection Using the Force–Displacement Reciprocity: Application to the Compact Compression Specimen, Int. J. Fract., **73** (1995), 67–79.

S. Mall and A.S. Kobayashi: Dynamic Fracture Toughness of Homalite-100, Exp. Mech., **18** (1978), 11–18.

B. Mandelbrot, D.E. Passoja and A.J. Paullay: Fractal Character of Fracture Surfaces in Metals, Nature, **308** (1984), 721–722.

P. Mannogg: Schattenoptische Messung der Spezifishen Burchenergie Während des Bruchvorgangs bei Plexiglas, Proceedings of the International Conference on the Physics of Non-crystalline Solids, Delft, The Netherlands, 1964, pp. 481–490.

M. Marder and S.P. Gross: Origin of Crack Tip Instabilities, J. Mech. Phys. Solids, **43** (1995), 1–48.

J.J. Mason, J. Lambros and A.J. Rosakis: The Use of Coherent Gradient Sensor in Dynamic Mixed-Mode Fracture Mechanics Experiments, J. Mech. Phys. Solids, **40** (1992), 641–661.

K. Matsushinge, Y. Sakurada and K. Takahashi: X-Ray Microanalysis and Acoustic Emission Studies on the Formation Mechanism of Secondary Cracks in PMMA, J. Mater. Sci., **19** (1984), 1548.

A.W. Maue: Die Beugung Elastischer Wellen an der Halbebene, Z. Angew. Math. Mech., **33** (1953), 1–10.

J. Miklowitz: The Theory of Elastic Waves and Waveguides, North-Holland Publishing Company, Amsterdam, 1978.

I. Milne, R.A. Ainsworth, A.R. Dowling and A.T. Stewart: Assessment of the Integrity of Structures Containing Defects, Int. J. Press. Vessels Piping, **32** (1988), 3–104.

J.W. Morrissey and J.R. Rice: Crack Front Waves, J. Mech. Phys. Solids, **46** (1998), 467–487.

J.W. Morrissey and J.R. Rice: Perturbative Simulations of Rack Front Waves, J. Mech. Phys. Solids, **48** (2000), 1229–1251.

N.F. Mott: Brittle Fracture in Mild Steel Plates, Engineering, **165** (1948), 16–18.

T. Nakamura, C.F. Shih and L.B. Freund: Analysis of a Dynamically Loaded Three Point Bend Ductile Fracture Specimen, Engng Fract. Mech., **25** (1986), 323–339.

A. Nakano, R.K. Kalia and P. Vashishta: Dynamics and Morphology of Brittle Cracks: A Molecular-Dynamics Study of Silicon Nitride, Phys. Rev. Lett., **75** (1995), 3138–3141.

T. Nicholas: Instrumented Impact Testing Using a Hopkinson Bar Apparatus. Technical Report, AFML-TR-7554, Wright-Patterson AFB, Ohio, 1975.

B. Noble: Methods Based on the Wiener-Hopf Technique, Pergamon Press, Elmsford, NY, 1958.

G. Nomarski: French Patent Specification 1,059,123, 1955.

P.D. O'Donoghue, M.F. Kanninen, C.P. Leung, G. Demofonti and S. Venzi: The Development and Validation of a Dynamic Fracture Propagation Model for Gas Transmission Pipelines, Int. J. Press. Vessels Piping, **70** (1997), 11–25.

M. Ortiz and A. Pandolfi: Finite-Deformation Irreversible Cohesive Elements for Three-Dimensional Crack Propagation Analysis, Int. J. Numer. Methods Engng, **44** (1999), 1267–1282.

D.M. Owen, S. Zhuang, A.J. Rosakis and G. Ravichandran: Experimental Determination of Dynamic Initiation and Propagation Fracture Toughness in Thin Aluminum Sheets, Int. J. Fract., **90** (1998), 153–174.

T.L. Paxson and R.A. Lucas: An Investigation of the Velocity Characteristics of a Fixed Boundary Fracture Model, In: G.C. Sih (Ed): Dynamic Crack Propagation, Noordhoff International Publishing, Leyden, 1973, pp. 415–426.

R.D. Pfaff, P.D. Washabaugh and W.G. Knauss: An Interpretation of Twyman–Green Interferograms from Static and Dynamic Fracture Experiments, Int. J. Solids Struct., **32** (1995), 939–956.

W.H. Press, S.A. Teukolsky, W.T. Vettering and B.P. Flannery: Numerical Recipes in C; The Art of Scientific Computing, 2nd ed., Cambridge University Press, Cambridge, 1992.

C.E. Pugh, D.J. Naus, B.R. Bass, R.K. Nanstad, R. deWit, R.J. Fields and S.R. Low, III: Wide-Plate Crack-Arrest Tests Utilizing Prototypical Pressure Vessel Steel, Int. J. Press. Vessels Piping, **31** (1988), 165–188.

S. Ramanathan and D.S. Fisher: Dynamics and Instabilities of Planar Tensile Cracks in Heterogeneous Media, Phys. Rev. Lett., **79** (1997), 877–880.

K. Ravi-Chandar: A Note on the Dynamic Stress Field Near a Propagating Crack, Int. J. Solids Struct., **19** (1983), 839–841.

K. Ravi-Chandar: On the Failure Mode Transitions in Polycarbonate Under Dynamic Mixed-Mode Loading, Int. J. Solids Struct., **32** (1995), 925–938.

K. Ravi-Chandar: Dynamic Shear Cracks Propagation in Homogeneous Materials, Proceedings of the 10th International Conference on Fracture, Elsevier, December 2001.

K. Ravi-Chandar and W.G. Knauss: Dynamic Crack Tip Stresses Under Stress Wave Loading—A Comparison of Theory and Experiment, Int. J. Fract., **20** (1982), 209–222.

K. Ravi-Chandar and W.G. Knauss: An Experimental Investigation into Dynamic Fracture—I. Crack Initiation and Crack Arrest, Int. J. Fract., **25** (1984a), 247–262.

K. Ravi-Chandar and W.G. Knauss: An Experimental Investigation into Dynamic Fracture—II. Microstructural Aspects, Int. J. Fract., **26** (1984b), 65–80.

K. Ravi-Chandar and W.G. Knauss: An Experimental Investigation into Dynamic Fracture—III. On Steady State Crack Propagation and Branching, Int. J. Fract., **26** (1984c), 141–154.

K. Ravi-Chandar and W.G. Knauss: An Experimental Investigation into Dynamic Fracture—IV. On the Interaction of Stress Waves with Propagating Cracks, Int. J. Fract., **26** (1984d), 189–200.

K. Ravi-Chandar and W.G. Knauss: On the Characterization of the Transient Stress Field Near the Tip of a Crack, J. Appl. Mech., **54** (1987), 72–78.

K. Ravi-Chandar and B. Yang: On the Role of Microcracks in the Dynamic Fracture of Brittle Materials, J. Mech. Phys. Solids, **45** (1997), 535–563.

G. Ravichandran and R.J. Clifton: Dynamic Fracture Under Plane Wave Loading, Int. J. Fract., **40** (1989), 157–201.

V.R. Regel: J. Tech. Phys. USSR, **21** (1951), 287.

J.R. Rice: Mathematical Analysis in the Mechanics of Fracture, In: H.L. Liebowitz (Ed.): Fracture, Vol. 2, Academic Press, New York, 1968, pp. 191–311.

J.R. Rice, P.C. Paris and J.G. Merkle: Some Further Results of j-Integral Analysis and Estimates, Progress in Flaw Growth and Fracture Toughness Testing, ASTM STP 536, American Society for Testing and Materials, Philadelphia, 1973, pp. 231–245.

H.G. Richter and F. Kerkhof: Stress Wave Fractography, In: R.C. Bradt and R.E. Tressler (Eds): Fractography in Glass, Plenum Press, New York, 1994, pp. 75–109.

W.F. Riley and J.W. Dally: Recording Dynamic Fringe Patterns with a Cranz-Schardin Camera, Exp. Mech., **9** (1969), 27N–33N.

D. Rittel and H. Maigre: An Investigation of Dynamic Crack Initiation in PMMA, Mech. Mater., **23** (1995), 229–239.

T.S. Robertson: Propagation of a Brittle Fracture in Steel, J. Iron Steel Inst., **175** (1953), 361–374.

A.J. Rosakis: Analysis of the Optical Method of Caustics for Dynamic Crack Propagation, Engng Fract. Mech., **13** (1980), 331–347.

A.J. Rosakis: Two Optical Techniques Sensitive to the Gradients of Optical Path Difference: The Method of Caustics and the Coherent Gradient Sensor, In: J.S. Epstein (Ed): Experimental Techniques in Fracture, Vol III, VCH Publishers, New York, 1993, pp. 327–425.

A.J. Rosakis: Intersonic Shear Cracks and Fault Ruptures, Adv. Phys., **51** (2002), 1189–1257.

A.J. Rosakis and K. Ravi-Chandar: On Crack Tip Stress State: An Experimental Evaluation of Three-Dimensional Effects, Int. J. Solids Struct., **22** (1986), 121–134.

A.J. Rosakis, J. Duffy and L.B. Freund: The Determination of Dynamic Fracture Toughness of AISI 4340 Steel by the Shadow Spot Method, J. Mech. Phys. Solids, **32** (1984), 443–460.

A.J. Rosakis, C. Liu and L.B. Freund: A Note on the Asymptotic Stress Field of a Non-uniformly Propagating Dynamic Crack, Int. J. Fract., **50** (1991), R39–R45.

A.J. Rosakis, R.P. Singh, Y. Tsuji, E. Kolawa and N.R. Moore, Jr.: Full Field Measurements of Curvature Using Coherent Gradient Sensing: Application to Thin Film Characterization, Thin Solid Films, **325** (1998), 42–54.

A.J. Rosakis, O. Samudrala and D. Coker: Cracks Faster than the Shear Wave Speed, Science, **284** (1999), 1337–1340.

C. Ruiz and R.A.W. Mines: The Hopkinson Pressure Bar: An Alternative to the Instrumented Pendulum for Charpy Tests, Int. J. Fract., **29** (1985), 101–109.

R.A. Schapery: A Theory of Crack Initiation and Growth in Viscoelastic Media. I. Theoretical Development, Int. J. Fract., **11** (1975), 141–159.

H. Schardin: Velocity Effects in Fracture, In: B.L. Averbach et al. (Eds): Fracture, Wiley, New York, 1959, pp. 297–330.

H. Schardin and W. Struth: Hochfrequenzkinematographische Untersuchung der Bruchvorgänge in Glas, Glastechnische Berichte, 1938, 219.

E. Sharon and J. Fineberg: Microbranching Instability and the Dynamic Fracture of Brittle Materials, Phys. Rev. B, **54** (1996), 7128–7139.

E. Sharon, G. Cohen and J. Fineberg: Propagating Solitary Waves Along a Rapidly Moving Crack Front, Nature, **410** (2001), 68–71.

D.A. Shockey and D.R. Curran: A Method for Measuring K_{IC} at Very High Strain Rates, Progress in Flaw Growth and Fracture Toughness Testing, ASTM STP 536, American Society for Testing and Materials, Philadelphia, 1973, pp. 297–311.

D.A. Shockey, J.F. Kalthoff, W. Klemm and S. Winkler: Simultaneous Measurements of Stress Intensity and Toughness for Fast-Running Crack in Steels, Exp. Mech., **23** (1983a), 140–145.

D.A. Shockey, J.F. Kalthoff and D.C. Erlich: Evaluation of Dynamic Crack Instability, Int. J. Fract., **22** (1983b), 217–229.

A. Shukla and H.P. Rossmanith: Dynamic Photoelastic Investigation of Wave Propagation and Energy Transfer Across Contacts, J. Strain Anal., **21** (1995), 213–218.

A. Shukla, R.K. Agarwal and H. Nigam: Dynamic Fracture Studies on 7075-T6 Aluminum and 4340 Steel Using Strain Gages and Photoelastic Coatings, Engng Fract. Mech., **31** (1989), 501–515.

L.I. Slepyan: Models and Phenomena in Fracture Mechanics, Springer, New York, 2002.

L.I. Slepyan and A.L. Fishkov: The Problem of the Propagation of a Cut at Transonic Velocity, Soviet Phys. Dokl., **26** (1981), 1192–1193.

E. Smekal: Zum Bruchvorgang bei Sprodem Stoffverhalten Unter ein- and Mehrachsigen Beanspruchungen, Osterreichische Ingenieur Arch., **7** (1953), 49–70.

I.N. Sneddon: The Distribution of Stress in the Neighborhood of a Crack in an Elastic Solid, Proc. R. Soc. Lond., **A187** (1946), 229–260.

J.E. Srawley: Wide Range Stress Intensity Factor Expressions for ASTM E399 Standard Fracture Toughness Specimens, Int. J. Fract., **12** (1976), 475–476.

L.S. Srinath: Scattered Light Photoelasticity, Tata McGraw Hill, New Delhi, 1983.

B. Stalder, P. Beguelin and H.H. Kausch: A Simple Velocity Gauge for Measuring Crack Growth, Int. J. Fract., **22** (1983), R47–R54.

E. Sternberg: On the Integration of the Equations of Motion in the Classical Theory of Elasticity, Arch. Rational Mech. Anal., **6** (1960).

G. Subramaninan and S.M. Nair: Direct Determiniation of Curvatures of Bent Plates Using a Double-Glass-Plate Shearing Interferometer, Exp. Mech., **25** (1985), 376–380.

H. Tada: The Stress Analysis of Cracks Handbook, Del Research Corporation, New York, 1973.

C. Taudou, S.V. Potti and K. Ravi-Chandar: On the Dominance of the Singular Dynamic Crack Tip Stress Field Under High Rate Loading, Int. J. Fract., **56** (1992), 41–59.

P.S. Theocaris: Localized Yielding Around a Crack Tip in Plexiglas, J. Appl. Mech., **37** (1970), 409–415.

R. Thompson: Lattice Trapping of Fracture Cracks, J. Appl. Phys., **42** (1971), 3154–3160.

H.V. Tippur, S. Krishnaswamy and A.J. Rosakis: A Coherent Gradient Sensor for Crack Tip Deformation Measurements: Analysis and Experimental Results, Int. J. Fract., **48** (1990), 193–204.

H.V. Tippur, S. Krishnaswamy and A.J. Rosakis: Optical Mapping of Crack Tip Deformations Using the Methods of Transmission and Reflection Coherent Gradient Sensing: A Study of Crack Tip K-Dominance, Int. J. Fract., **52** (1991), 91–117.

UK Air Accidents Investigation Branch: Report on the Accident to Boeing 747-121, N739PA at Lockerbie, Dumfriesshire, Scotland on 21 December, 1988, Aircraft Accident Report No 2/90 (EW/C1094), 1990.

F.F. Videon, F.W. Barton and W.J. Hall: Brittle Fracture Propagation Studies, Ship Structure Committee Report SSC-148, Aug 1963, Office of Technical Services, Washington DC, 1963.

I.A. Viktorov: Rayleigh and Lamb Waves: Physical Theory and Applications, Plenum Press, New York, 1967.

W.B. Wade and A.S. Kobayashi: An Investigation of Propagating Cracks by Dynamic Photoelasticity, Exp. Mech., **10** (1970), 106–113.

W. Waetzmann: Nature, **172** (1912), 461.

H. Wallner: Linenstrukturen an Bruchflächen, Z. Phys., **114** (1939), 368–378.

P.D. Washabaugh and W.G. Knauss: Nonsteady Periodic Behavior in the Dynamic Fracture of PMMA, Int. J. Fract., **59** (1993), 189.

L.T. Wheeler and E. Sternberg: Some Theorems in Classical Elastodynamics, Arch. Rational Mech. Anal., **31** (1968), 51–90.

M.L. Williams: On the Stress Distribution at the Base of a Stationary Crack, J. Appl. Mech., **24** (1957) 109–114.

J.R. Willis: Equations of Motion for Propagating Cracks, The Mechanics and Physics of Fracture, The Metals Society, 1975, pp. 57–67.

J.R. Willis: Accelerating Cracks and Related Problems, In: G. Eason and R.W. Ogden (Eds): Elasticity, Mathematical Methods and Applications, The Ian Sneddon 70th Birthday Volume, Ellis Horwood, Chichester, 1989, pp. 397–409.

J.R. Willis: The Stress Field Near the Tip of an Accelerating Crack, J. Mech. Phys. Solids, **40** (1992), 1671–1681.

M.L. Wilson, R.H. Hawley and J. Duffy: The Effect of Loading Rate and Temperature on Fracture Initiation in 1020 Hot-Rolled Steel, Engng Fract. Mech., **13** (1980), 371–385.

X.-P. Xu and A. Needleman: Numerical Simulations of Fast Crack Growth in Brittle Solids, J. Mech. Phys. Solids, **42** (1994), 1397–1434.

B. Yang and K. Ravi-Chandar: On the Role of the Process Zone in Dynamic Fracture, J. Mech. Phys. Solids, **44** (1996), 1955–1976.

E. Yoffe: The Moving Griffith Crack, Philos. Mag., **42** (1951), 739–750.

S.N. Zhurkov: Kinetic Concept of Strength of Solids, Int. J. Fract., **1** (1965), 311–323.

Further Reading

H. Coque, C.P. Leung and S.H. Hudak, Jr.: Effect of Planar Size and Dynamic Loading Rate on Initiation and Propagation Toughness of a Moderate-Roughness Steel, Engng Fract. Mech., **47** (1994), 249–267.

F. Costanzo and J.R. Walton: Numerical Simulations of a Dynamically Propagating Crack with a Nonlinear Cohesive Zone, Int. J. Fracture, **91** (1998), 373–389.

L.S. Costin and J. Duffy: The Effect of Loading Rate and Temperature on the Initiation of Fracture in Mild, Rate-Sensitive Steel, J. Engng Mater. Technol., **101** (1979), 258–264.

J.W. Dally and A. Shukla: Dynamic Crack Behavior at Initiation, Mech. Res. Commun., **6** (1979), 239–244.

J.W. Dally and A. Shukla: Influence of Late-Breaking Ligaments on Crack-Propagation in Compact Specimens – A Photo-Elastic Study, Exp. Mech., **23** (1983), 298–303.

A.T. de Hoop: Representation Theorems for the Displacement in an Elastic Solid and Their Application to Elastodynamic Diffraction Theory, Ph.D. Thesis, Technical University, Delft, 1958, p. 72.

L.B. Freund: Crack Propagation in an Elastic Solid Subjected to General Loading. II. Nonuniform Rate of Extension, J. Mech. Phys. Solids, **20** (1972c), 141–150.

S.K. Khanna and A. Shukla: On the Use of Strain Gages in Dynamic Fracture Mechanics, Engng Fract. Mech., **51** (1995), 933–948.

W.G. Knauss and K. Ravi-Chandar: Some Basic Problems in Stress Wave Dominated Fracture, Int. J. Fracture, **27** (1985), 127–143.

A.S. Kobayashi: Handbook on Experimental Mechanics, Prentice Hall, New York, 1993.

T. Kobayashi, I. Yamamoto and M. Niinomi: Evaluation of the Dynamic Toughness Parameters by Instrumented Charpy Impact Test, Engng Fract. Mech., **24** (1986), 773–782.

T.L. Leise and J.R. Walton: A Method for Solving Dynamically Accelerating Crack Problems in Linear Viscoelasticity, SIAM J. Applied Math., **64** (2003), 94–107.

J.J. Mason, A.J. Rosakis and G. Ravichandran: Full Field Measurements of the Dynamic Deformation Field Around a Growing Adiabatic Shear Band at the Tip of a Dynamically Loaded Crack or Notch, J. Mech. Phys. Solids, **42** (1994), 1679–1697.

K. Ravi-Chandar: An Experimental Investigation into the Mechanics of Dynamic Fracture, Ph.D. Thesis, California Institute of Technology, Pasadena, 1982.

J.M. Rolfe and S.T. Barsom: Correlations Between K_{IC} and Charpy V-Notch Test Results in the Transition Temperature Range, Impact Testing of Metals, ASTM STP 466, American Society for Testing and Materials, Philadelphia, 1970, pp. 281–302.

A. Shukla and H. Nigam: A Note on the Stress Intensity Factor and Crack Velocity Relationship for Homalite-100, Engng Fracture Mech, **25** (1986), 91.

W.B. Wade and A.S. Kobayashi: Photoelastic Investigation on the Crack-Arrest Capability of a Pretensioned Stiffened Plate, Exp. Mech., **15** (1975), 1–9.

Appendix A

Dynamic Crack Tip Asymptotic Fields

A1 Dynamic Crack Tip Stress Field for a Stationary Crack

For a stationary crack loaded with a time-dependent load, the stress field in the vicinity of the crack tip is characterized through the following equation

$$\sigma_{\alpha\beta}(r, \theta) = \frac{K_{\mathrm{I}}(t)}{\sqrt{2\pi r}} f_{\alpha\beta}^{\mathrm{Is}}(\theta) + \frac{K_{\mathrm{II}}(t)}{\sqrt{2\pi r}} f_{\alpha\beta}^{\mathrm{IIs}}(\theta) + \sigma_{0x}\delta_{\alpha 1}\delta_{\beta 1} + \cdots \text{ as } r \to 0 \qquad (A1.1)$$

where $K_{\mathrm{I}}(t)$ and $K_{\mathrm{II}}(t)$ are the mode I and mode II dynamic stress intensity factors and σ_{0x} is the first nonsingular term in the asymptotic expansion. The angular variation of the functions $f_{\alpha\beta}^{\mathrm{Is}}(\theta)$ and $f_{\alpha\beta}^{\mathrm{IIs}}(\theta)$ are given below:

$$\begin{aligned}
f_{11}^{\mathrm{Is}}(\theta) &= \cos\tfrac{1}{2}\theta[1 - \sin\tfrac{1}{2}\theta\sin\tfrac{3}{2}\theta], \\
f_{22}^{\mathrm{Is}}(\theta) &= \cos\tfrac{1}{2}\theta[1 + \sin\tfrac{1}{2}\theta\sin\tfrac{3}{2}\theta], \qquad\qquad (A1.2) \\
f_{12}^{\mathrm{Is}}(\theta) &= \cos\tfrac{1}{2}\theta\sin\tfrac{1}{2}\theta\cos\tfrac{3}{2}\theta
\end{aligned}$$

$$\begin{aligned}
f_{11}^{\mathrm{IIs}}(\theta) &= -\sin\tfrac{1}{2}\theta[2 + \cos\tfrac{1}{2}\theta\cos\tfrac{3}{2}\theta], \\
f_{22}^{\mathrm{IIs}}(\theta) &= \cos\tfrac{1}{2}\theta\sin\tfrac{1}{2}\theta\cos\tfrac{3}{2}\theta, \qquad\qquad (A1.3) \\
f_{12}^{\mathrm{IIs}}(\theta) &= \cos\tfrac{1}{2}\theta[1 - \sin\tfrac{1}{2}\theta, \sin\tfrac{3}{2}\theta]
\end{aligned}$$

A2 Steady-State Dynamic Crack Tip Stress Field: Singular Term

For easy reference, the asymptotic dynamic crack tip stress field is written down here for different cases. The general form of the steady-state stress field is given by

$$\begin{aligned}
\sigma_{\alpha\beta}(r, \theta) &= \frac{K_{\mathrm{I}}^{\mathrm{dyn}}}{\sqrt{2\pi r}} f_{\alpha\beta}^{\mathrm{I}}(\theta; v) + \frac{K_{\mathrm{II}}^{\mathrm{dyn}}}{\sqrt{2\pi r}} f_{\alpha\beta}^{\mathrm{II}}(\theta; v) + \sigma_{0x}\delta_{\alpha 1}\delta_{\beta 1} + \cdots \text{ as } r \to 0, \\
u_{\alpha}(r, \theta) &= \frac{K_{\mathrm{I}}^{\mathrm{dyn}}\sqrt{r}}{\mu\sqrt{2\pi}} g_{\alpha}^{\mathrm{I}}(\theta; v) + \frac{K_{\mathrm{II}}^{\mathrm{dyn}}}{\mu\sqrt{2\pi r}} g_{\alpha}^{\mathrm{II}}(\theta; v) + \cdots \text{ as } r \to 0 \qquad (A2.1)
\end{aligned}$$

where $K_{\mathrm{I}}^{\mathrm{dyn}}$ and $K_{\mathrm{II}}^{\mathrm{dyn}}$ are the mode I and mode II dynamic stress intensity factors and σ_{0x} is the first nonsingular term in the asymptotic expansion. The angular variation of the functions $f^{\mathrm{I}}_{\alpha\beta}(\theta; v), f^{\mathrm{II}}_{\alpha\beta}(\theta; v), g^{\mathrm{I}}_{\alpha\beta}(\theta; v)$, and $g^{\mathrm{II}}_{\alpha\beta}(\theta; v)$ are given below

$$f^{\mathrm{I}}_{11}(\theta; v) = \frac{1}{R(v)}\left\{(1+\alpha_{\mathrm{s}}^2)(1+2\alpha_{\mathrm{d}}^2 - \alpha_{\mathrm{s}}^2)\frac{\cos\frac{1}{2}\theta_{\mathrm{d}}}{\gamma_{\mathrm{d}}^{1/2}} - 4\alpha_{\mathrm{d}}\alpha_{\mathrm{s}}\frac{1}{\gamma_{\mathrm{s}}^{1/2}}\cos\tfrac{1}{2}\theta_{\mathrm{s}}\right\},$$

$$f^{\mathrm{I}}_{22}(\theta; v) = -\frac{1}{R(v)}\left\{(1+\alpha_{\mathrm{s}}^2)^2\frac{\cos\frac{1}{2}\theta_{\mathrm{d}}}{\gamma_{\mathrm{d}}^{1/2}} - 4\alpha_{\mathrm{d}}\alpha_{\mathrm{s}}\frac{1}{\gamma_{\mathrm{s}}^{1/2}}\cos\tfrac{1}{2}\theta_{\mathrm{s}}\right\}, \tag{A2.2}$$

$$f^{\mathrm{I}}_{12}(\theta; v) = \frac{2\alpha_{\mathrm{d}}(1+\alpha_{\mathrm{s}}^2)}{R(v)}\left\{\frac{\sin\frac{1}{2}\theta_{\mathrm{d}}}{\gamma_{\mathrm{d}}^{1/2}} - \frac{\sin\frac{1}{2}\theta_{\mathrm{s}}}{\gamma_{\mathrm{s}}^{1/2}}\right\}$$

$$f^{\mathrm{II}}_{11}(\theta; v) = -\frac{2\alpha_{\mathrm{s}}}{R(v)}\left\{(1+2\alpha_{\mathrm{d}}^2 - \alpha_{\mathrm{s}}^2)\frac{\sin\frac{1}{2}\theta_{\mathrm{d}}}{\gamma_{\mathrm{d}}^{1/2}} - (1+\alpha_{\mathrm{s}}^2)\frac{1}{\gamma_{\mathrm{s}}^{1/2}}\sin\tfrac{1}{2}\theta_{\mathrm{s}}\right\},$$

$$f^{\mathrm{II}}_{22}(\theta; v) = \frac{2\alpha_{\mathrm{s}}(1+\alpha_{\mathrm{s}}^2)}{R(v)}\left\{\frac{\sin\frac{1}{2}\theta_{\mathrm{d}}}{\gamma_{\mathrm{d}}^{1/2}} - \frac{1}{\gamma_{\mathrm{s}}^{1/2}}\sin\tfrac{1}{2}\theta_{\mathrm{s}}\right\}, \tag{A2.3}$$

$$f^{\mathrm{II}}_{12}(\theta; v) = \frac{1}{R(v)}\left\{4\alpha_{\mathrm{d}}\alpha_{\mathrm{s}}\frac{\cos\frac{1}{2}\theta_{\mathrm{d}}}{\gamma_{\mathrm{d}}^{1/2}} - (1+\alpha_{\mathrm{s}}^2)\frac{\cos\frac{1}{2}\theta_{\mathrm{s}}}{\gamma_{\mathrm{s}}^{1/2}}\right\}$$

$$g^{\mathrm{I}}_1(r, \theta) = \frac{2}{R(v)}\left\{(1+\alpha_{\mathrm{s}}^2)r_{\mathrm{d}}^{1/2}\cos\left(\frac{\theta_{\mathrm{d}}}{2}\right) - 2\alpha_{\mathrm{d}}\alpha_{\mathrm{s}}r_{\mathrm{s}}^{1/2}\cos\left(\frac{\theta_{\mathrm{s}}}{2}\right)\right\},$$

$$\tag{A2.4}$$

$$g^{\mathrm{I}}_2(r, \theta) = -\frac{2\alpha_{\mathrm{d}}}{R(v)}\left\{(1+\alpha_{\mathrm{s}}^2)r_{\mathrm{d}}^{1/2}\sin\left(\frac{\theta_{\mathrm{d}}}{2}\right) - 2r_{\mathrm{s}}^{1/2}\sin\left(\frac{\theta_{\mathrm{s}}}{2}\right)\right\}$$

$$g^{\mathrm{II}}_1(\theta, v) = \frac{2\alpha_{\mathrm{s}}}{R(v)}\left\{2r_{\mathrm{d}}^{1/2}\sin\left(\frac{\theta_{\mathrm{d}}}{2}\right) - (1+\alpha_{\mathrm{s}}^2)r_{\mathrm{s}}^{1/2}\sin\left(\frac{\theta_{\mathrm{s}}}{2}\right)\right\},$$

$$\tag{A2.5}$$

$$g^{\mathrm{II}}_2(\theta, v) = \frac{2}{R(v)}\left\{2\alpha_{\mathrm{d}}\alpha_{\mathrm{s}}r_{\mathrm{d}}^{1/2}\sin\left(\frac{\theta_{\mathrm{d}}}{2}\right) - (1+\alpha_{\mathrm{s}}^2)r_{\mathrm{s}}^{1/2}\sin\left(\frac{\theta_{\mathrm{s}}}{2}\right)\right\}$$

$$r_{\mathrm{d}} = \sqrt{x_1^2 + \alpha_{\mathrm{d}}^2 x_2^2}, \quad \tan\theta_{\mathrm{d}} = \tan\left(\frac{\alpha_{\mathrm{d}}x_2}{x_1}\right) \tag{A2.6}$$

$$r_{\mathrm{s}} = \sqrt{x_1^2 + \alpha_{\mathrm{s}}^2 x_2^2}, \quad \tan\theta_{\mathrm{s}} = \tan\left(\frac{\alpha_{\mathrm{s}}x_2}{x_1}\right) \tag{A2.7}$$

$$\alpha_{\mathrm{d}} = \sqrt{1 - \frac{v^2}{C_{\mathrm{d}}^2}}, \quad \alpha_{\mathrm{s}} = \sqrt{1 - \frac{v^2}{C_{\mathrm{s}}^2}} \tag{A2.8}$$

$$R(v) = 4\alpha_{\mathrm{d}}\alpha_{\mathrm{s}} - (1+\alpha_{\mathrm{s}}^2)^2 \tag{A2.9}$$

C_d is the dilatational wave speed and C_s the shear wave speed. For thin plates, the appropriate dilatational wave speed is the plate wave speed given by:

$$C_d^p = \sqrt{\frac{E}{\rho(1 - \nu^2)}} \tag{A2.10}$$

A3 Steady-State Crack Tip Displacement and Stress Field: N Terms

The steady-state crack tip displacement and stress components for symmetric loading (mode I) about the crack line are given below:

$$
\begin{aligned}
u_1^n(r, \theta) &= A_n\left(1 + \frac{n}{2}\right)\left\{r_d^{n/2} \cos\left(\frac{n\theta_d}{2}\right) - \chi_1^1(n)r_s^{n/2} \cos\left(\frac{n\theta_s}{2}\right)\right\}, \\
u_2^n(r, \theta) &= A_n\alpha_d\left(1 + \frac{n}{2}\right)\left\{-r_d^{n/2} \sin\left(\frac{n\theta_d}{2}\right) + \chi_2^1(n)r_s^{n/2} \sin\left(\frac{n\theta_s}{2}\right)\right\}
\end{aligned}
\tag{A3.1}
$$

$$
\begin{aligned}
\varepsilon_{11}^n(r, \theta) &= A_n\frac{n}{2}\left(1 + \frac{n}{2}\right)\left\{r_d^{n/2-1} \cos\left(\frac{n-2}{2}\theta_d\right) - \chi_1^1(n)r_s^{n/2-1} \cos\left(\frac{n-2}{2}\theta_s\right)\right\}, \\
\varepsilon_{22}^n(r, \theta) &= A_n\frac{n}{2}\left(1 + \frac{n}{2}\right)\left\{-\alpha_d^2 r_d^{n/2-1} \cos\left(\frac{n-2}{2}\theta_d\right)\right. \\
&\qquad\qquad\qquad\qquad\left. + \chi_1^1(n)r_s^{n/2-1} \cos\left(\frac{n-2}{2}\theta_s\right)\right\}, \\
\varepsilon_{12}^n(r, \theta) &= \frac{A_n}{4\alpha_s}\frac{n}{2}\left(1 + \frac{n}{2}\right)\left\{-4\alpha_d\alpha_s r_d^{n/2-1} \sin\left(\frac{n-2}{2}\theta_d\right)\right. \\
&\qquad\qquad\qquad\qquad\left. + \kappa_1(n)r_s^{n/2-1} \sin\left(\frac{n-2}{2}\theta_s\right)\right\}
\end{aligned}
\tag{A3.2}
$$

$$
\begin{aligned}
\sigma_{11}^n(r, \theta) &= \frac{\mu A_n}{(1 + \alpha_s^2)}\frac{n}{2}\left(1 + \frac{n}{2}\right)\left\{(1 + \alpha_s^2)(1 + 2\alpha_d^2 - \alpha_s^2)r_d^{n/2-1} \cos\left(\frac{n-2}{2}\theta_d\right)\right. \\
&\qquad\qquad\qquad\qquad\left. - \kappa_1(n)r_s^{n/2-1} \cos\left(\frac{n-2}{2}\theta_s\right)\right\}, \\
\sigma_{22}^n(r, \theta) &= \frac{\mu A_n}{(1 + \alpha_s^2)}\frac{n}{2}\left(1 + \frac{n}{2}\right)\left\{-(1 + \alpha_s^2)^2 r_d^{n/2-1} \cos\left(\frac{n-2}{2}\theta_d\right)\right. \\
&\qquad\qquad\qquad\qquad\left. + \kappa_1(n)r_s^{n/2-1} \cos\left(\frac{n-2}{2}\theta_s\right)\right\}, \\
\sigma_{12}^n(r, \theta) &= \frac{\mu A_n}{2\alpha_s}\frac{n}{2}\left(1 + \frac{n}{2}\right)\left\{-4\alpha_d\alpha_s r_d^{n/2-1} \sin\left(\frac{n-2}{2}\theta_d\right)\right. \\
&\qquad\qquad\qquad\qquad\left. + \kappa_1(n)r_s^{n/2-1} \sin\left(\frac{n-2}{2}\theta_s\right)\right\}
\end{aligned}
\tag{A3.3}
$$

where

$$\chi_1^I(n) = \begin{cases} \dfrac{2\alpha_d\alpha_s}{1+\alpha_s^2}, & \text{for } n \text{ odd} \\[2ex] \dfrac{1+\alpha_s^2}{2}, & \text{for } n \text{ even} \end{cases} \quad \text{and} \quad \chi_2^I(n) = \begin{cases} \dfrac{2}{1+\alpha_s^2}, & \text{for } n \text{ odd} \\[2ex] \dfrac{1+\alpha_s^2}{2\alpha_d\alpha_s}, & \text{for } n \text{ even} \end{cases} \tag{A3.4}$$

$$\kappa_I(n) = \begin{cases} 4\alpha_d\alpha_s & \text{for } n \text{ odd} \\ (1+\alpha_s^2)^2, & \text{for } n \text{ even} \end{cases} \tag{A3.5}$$

For the case of antisymmetry (mode II), the displacement and stress components are

$$u_1^n(r,\theta) = A_n\left(1+\frac{n}{2}\right)\left\{r_d^{n/2}\sin\left(\frac{n\theta_d}{2}\right) - \chi_1^{II}(n)r_s^{n/2}\sin\left(\frac{n\theta_s}{2}\right)\right\},$$

$$u_2^n(r,\theta) = \alpha_d A_n\left(1+\frac{n}{2}\right)\left\{r_d^{n/2}\cos\left(\frac{n\theta_d}{2}\right) - \chi_2^{II}(n)r_s^{n/2}\cos\left(\frac{n\theta_s}{2}\right)\right\} \tag{A3.6}$$

$$\varepsilon_{11}^n(r,\theta) = A_n\frac{n}{2}\left(1+\frac{n}{2}\right)\left\{r_d^{n/2-1}\sin\left(\frac{n-2}{2}\theta_d\right) - \chi_1^{II}(n)r_s^{n/2-1}\sin\left(\frac{n-2}{2}\theta_s\right)\right\},$$

$$\varepsilon_{22}^n(r,\theta) = A_n\frac{n}{2}\left(1+\frac{n}{2}\right)\left\{-\alpha_d^2 r_d^{n/2-1}\sin\left(\frac{n-2}{2}\theta_d\right)\right.$$
$$\left. +\chi_1^{II}(n)r_s^{n/2-1}\sin\left(\frac{n-2}{2}\theta_s\right)\right\}, \tag{A3.7}$$

$$\varepsilon_{12}^n(r,\theta) = \frac{A_n}{4\alpha_s}\frac{n}{2}\left(1+\frac{n}{2}\right)\left\{4\alpha_d\alpha_s r_d^{n/2-1}\cos\left(\frac{n-2}{2}\theta_d\right)\right.$$
$$\left. -\kappa_{II}(n)r_s^{n/2-1}\cos\left(\frac{n-2}{2}\theta_s\right)\right\}$$

$$\sigma_{11}^n(r,\theta) = \frac{\mu A_n}{(1+\alpha_s^2)}\frac{n}{2}\left(1+\frac{n}{2}\right)\left\{(1+\alpha_s^2)(1+2\alpha_d^2-\alpha_s^2)r_d^{n/2-1}\sin\left(\frac{n-2}{2}\theta_d\right)\right.$$
$$\left. -\kappa_{II}(n)r_s^{n/2-1}\sin\left(\frac{n-2}{2}\theta_s\right)\right\},$$

$$\sigma_{22}^n(r,\theta) = \frac{\mu A_n}{(1+\alpha_s^2)}\frac{n}{2}\left(1+\frac{n}{2}\right)\left\{-(1+\alpha_s^2)^2 r_d^{n/2-1}\sin\left(\frac{n-2}{2}\theta_d\right)\right.$$
$$\left. +\kappa_{II}(n)r_s^{n/2-1}\sin\left(\frac{n-2}{2}\theta_s\right)\right\}, \tag{A3.8}$$

$$\sigma_{12}^n(r,\theta) = \frac{\mu A_n}{2\alpha_s}\frac{n}{2}\left(1+\frac{n}{2}\right)\left\{4\alpha_d\alpha_s r_d^{n/2-1}\cos\left(\frac{n-2}{2}\theta_d\right)\right.$$
$$\left. -\kappa_{II}(n)r_s^{n/2-1}\cos\left(\frac{n-2}{2}\theta_s\right)\right\}$$

where

$$\chi_1^{II}(n) = \begin{cases} \dfrac{2\alpha_d\alpha_s}{1+\alpha_s^2}, & \text{for } n \text{ even} \\ \dfrac{1+\alpha_s^2}{2}, & \text{for } n \text{ odd} \end{cases} \quad \text{and} \quad \chi_2^{II}(n) = \begin{cases} \dfrac{2}{1+\alpha_s^2}, & \text{for } n \text{ even} \\ \dfrac{1+\alpha_s^2}{2\alpha_d\alpha_s}, & \text{for } n \text{ odd} \end{cases} \tag{A3.9}$$

$$\kappa_{II}(n) = \begin{cases} 4\alpha_d\alpha_s, & \text{for } n \text{ even} \\ (1+\alpha_s^2)^2, & \text{for } n \text{ odd} \end{cases} \tag{A3.10}$$

A4 Transient Crack Tip Displacement and Stress Field: Six Terms

$$\frac{\sigma_{11}+\sigma_{22}}{2\rho(C_d^2-C_s^2)} = \frac{3v^2}{4C_d^2}A_0\cos\left(\frac{\theta_d}{2}\right)r_d^{-1/2} + \frac{2v^2}{c_d^2}A_1$$

$$+\left\{\left[\frac{15v^2}{4C_d^2}A_2 + \left(1-\frac{v^2}{2C_d^2}\right)D^1(A_0)\right]\cos\left(\frac{\theta_d}{2}\right)\right.$$

$$\left. +\frac{v^2}{8C_d^2}D^1(A_0)\cos\frac{3\theta_d}{2}\right\}r_d^{1/2}$$

$$+\left\{\left[\frac{6v^2}{C_d^2}A_3 + \left(1-\frac{v^2}{4C_d^2}\right)D^1(A_1)\right]\cos\theta_d\right\}r_d$$

$$+\left\{\left[\frac{35v^2}{4C_d^2}A_4 + \left(1-\frac{v^2}{2C_d^2}\right)D^1(A_2) + \frac{1}{9}\left(1-\frac{v^2}{4C_d^2}\right)D^2(A_0)\right.\right.$$

$$\left. +\left(1-\frac{v^2}{2C_d^2}\right)\ddot{A}_0\right]\cos\left(\frac{3\theta_d}{2}\right) + \left[\frac{3v^2}{8C_d^2}D^1(A_2)\right.$$

$$\left. +\frac{1}{6}\left(1-\frac{v^2}{4C_d^2}\right)D^2(A_0) + \frac{3v^2}{8C_d^2}\ddot{A}_0\right]\cos\left(\frac{\theta_d}{2}\right)$$

$$\left. +\left(\frac{v^2}{96C_d^2}D^2(A_0)\right)\cos\left(\frac{5\theta_d}{2}\right)\right\}r_d^{3/2}$$

$$+\left\{\left[\frac{12v^2}{C_d^2}A_5 + \left(1-\frac{v^2}{2C_d^2}\right)D^1(A_3) + \frac{D^2(A_1)}{16}\right.\right.$$

$$\left. +\left(1-\frac{v^2}{2C_d^2}\right)\ddot{A}_1\right]\cos(2\theta_d) + \left[\frac{v^2}{2C_d^2}D^1(A_3)\right.$$

$$\left. +\frac{1}{8}\left(1-\frac{v^2}{4C_d^2}\right)D^2(A_1) + \frac{v^2}{2C_d^2}\ddot{A}_1\right]\right\}r_d^2 + o(r_d^2),$$

$$\tag{A4.1}$$

where

$$D^1(A_k) = -\frac{(k+3)v}{C_d^2 \alpha_d^2} \frac{\mathrm{d}}{\mathrm{d}t}(A_k), \quad k = 0, 1, 2, \ldots$$

$$D^2(A_k) = D^1[D^1(A_k)] \quad \text{and} \quad \ddot{A}_k = \frac{1}{C_d^2 \alpha_d^2} \frac{\mathrm{d}^2}{\mathrm{d}t^2} A_k$$

Appendix B

Mechanical and Optical Properties of Selected Materials

Tables B.1–B.3.

Table B.1 Homalite-100 (reproduced from Ravi-Chandar and Knauss, 1982)

Modulus of elasticity (dynamic)	E	4550 MPa
Poisson's ratio	ν	0.31
Density	ρ	1230 kg/m^3
Plate wave speed	C_d^p	2057 m/s
Shear wave speed	C_s	1176 m/s
Rayleigh wave speed	C_R	1081 m/s
Index of refraction	n	1.5
Direct stress-optic coefficient	C_1	-0.906×10^{-10} m^2/N
Transverse stress-optic coefficient	C_2	-1.140×10^{-10} m^2/N
Plane strain fracture toughness	K_{IC}	0.44 MPa m$^{1/2}$

Table B.2 Polymethylmethacrylate (reproduced from Kalthoff, 1987)

Modulus of elasticity (dynamic)	E	3240 MPa
Poisson's ratio	ν	0.35
Density	ρ	1230 kg/m^3
Plate wave speed	C_d^p	1816 m/s
Shear wave speed	C_s	1035 m/s
Rayleigh wave speed	C_R	967 m/s
Index of refraction	n	1.491
Direct stress-optic coefficient	C_1	-0.530×10^{-10} m^2/N
Transverse stress-optic coefficient	C_2	-0.570×10^{-10} m^2/N
Plane strain fracture toughness	K_{IC}	1.05 MPa m$^{1/2}$

Table B.3 Araldite B (reproduced from Kalthoff, 1987)

Modulus of elasticity (dynamic)	E	3660 MPa
Poisson's ratio	ν	0.392
Density	ρ	1230 kg/m^3
Plate wave speed	C_d^p	1931 m/s
Shear wave speed	C_s	1065 m/s
Rayleigh wave speed	C_R	1001 m/s
Index of refraction	n	1.592
Direct stress-optic coefficient	C_1	-0.056×10^{-10} m^2/N
Transverse stress-optic coefficient	C_2	-0.620×10^{-10} m^2/N
Plane strain fracture toughness	K_{IC}	1.05 MPa m$^{1/2}$

Index

A

B

C